Bats

of Southern Africa

Bats
of
Southern Africa

Guide to Biology,
Identification, and Conservation

PETER JOHN TAYLOR

University of Natal Press
Pietermaritzburg
2000

Published by University of Natal Press
Private Bag X01
Scottsville
3209
South Africa

ISBN 0 86980 982 2

Cover design by Brett Armstrong

Generous donations were received from Endangered Wildlife Trust (EWT), Linton Projects (Pty) Ltd, and Klatzko-Waldrin Advertising.

Typeset by Alpha Typesetters cc, New Germany
Printed by Teeanem Printers (Pty) Ltd
208 Victoria Road, Pietermaritzburg 3201

Contents

Foreword

Dr Peter Taylor has served as Curator of Mammals at the Durban Natural Science Museum since 1989, and in that capacity has not only become an expert on African bats, but also a leader in public education and conservation efforts. He founded the Durban Bat Interest Group in 1994 and subsequently has played a key role in advancing bat conservation throughout southern Africa. In *Bats of Southern Africa,* he provides the first of its kind introduction of bats to the public, as well as an invaluable summary of current knowledge for colleagues in bat research.

As you will learn from this book, bats rank among Africa's most fascinating and important animals. They are the only mammals that fly, and their orientation systems are a scientific marvel. Many bats have ears so sensitive that they can hear the footsteps of small insects. Others have suction discs for climbing on slick leaf surfaces, spectacular crests and epaulets for courtship display, brilliant colors and designs for camouflage and attracting mates, and much, much more.

Nearly a thousand kinds of bats comprise almost a quarter of all mammal species, and they live nearly everywhere. As you will learn, bats play essential roles in maintaining the balance of nature and human economies. Those that eat insects are primary predators of the vast numbers that fly at night, including many of the world's most costly pests. Those that pollinate flowers and disperse seeds are essential to the maintenance of whole ecosystems, and bat-dependent plants produce crops valued in the hundreds of millions of dollars annually in cash-poor developing countries.

Despite the fact that bats rank among the world's most interesting and important mammals, they remain the most scientifically neglected, frequently misunderstood and needlessly persecuted animals. Because many species produce only a single young each year and live in conspicuous colonies numbering from thousands to millions, they are exceptionally vulnerable to extinction. Like other animals, they are increasingly threatened by habitat loss and pollution, but for many bats, the single most important threat comes from people who kill them in large numbers out of fear and ignorance.

In *Bats of Southern Africa,* Dr Taylor has done an outstanding job sharing his personal enthusiasm about bats while debunking age-old myths and introducing long neglected truths about these essential animals. For the layperson, this book will replace fear with admiration, opening a whole new world of wonder. For students, it will provide career ideas for lifetimes of work in a vast and exciting frontier of exploration, and for colleagues it will become a ready reference and a wonderful tool for sharing their enthusiasm about the little known world of bats.

Merlin D. Tuttle
Founder and Executive Director
Bat Conservation International

Acknowledgements

This book has become a reality through both generous sponsorships and the assistance and encouragement of friends, family and colleagues. The artwork by Christeen Grant was made possible through the financial contribution of the sponsors, Endangered Wildlife Trust (EWT), Linton Projects (Pty) Ltd, and Klatzko-Waldrin Advertising. Dr Merlin D. Tuttle and Bat Conservation International (BCI) kindly donated the colour photographs.

A number of colleagues assisted with reading and commenting on various drafts of the manuscript (in alphabetical order): Dr Gary Bronner, Fenton (Woody) Cotterill, Denis Eckard, Dr Brock Fenton, Nigel Fernsby, Teresa Kearney, Dr Ara Monadjem, Kate Richardson, and Dr Merlin D. Tuttle. I am indebted to Dr David Rowe-Rowe for critical scientific editing and proof-reading of the final draft, which led to considerable improvements of style and grammar. Of course any errors remain my own responsibility.

I am also very grateful to the curators of the mammal collections at the National Museum of Natural History in Paris (Professor Christiane Denys) and the Harrison Zoological Museum Institute (Drs David Harrison and Paul Bates) for allowing me access to study their valuable collections of African bats during a sabbatical visit in May and June 2000. These collections provided a valuable source of information on the description, identification, and measurements of various species of southern African bats not available in the collection of the Durban Natural Science Museum. I thank Willem Hechter and Dr John Bingham from Onderstepoort Veterinary Institute for advice on rabies issues, and Dr David Harrison for valuable discussions and advice on taxonomic problems of southern African bats. I am indebted to Woody Cotterill for generously sharing field notes and distributional data.

This book would not have been possible without the support and encouragement of the Director, staff and Friends of the Durban Natural Science Museum and enthusiastic members of the Durban Bat Interest Group, Gauteng Bat Interest Group, Cape Bat Action Team and other bat lovers, including Martijn Boonman who taught me much about bat detectors. I am grateful to all at the University of Natal Press for their commitment to this book and their professionalism, and to the artist, Christeen Grant, for her whole-hearted passion for her subject which truly showed itself in the resulting artwork. May her bats smile at you as they did for me.

Finally, I thank my precious family for patiently tolerating my virtual absence during the writing of this book over many evenings, weekends and weeks of 'leave' at home.

Bats Need Friends Too!

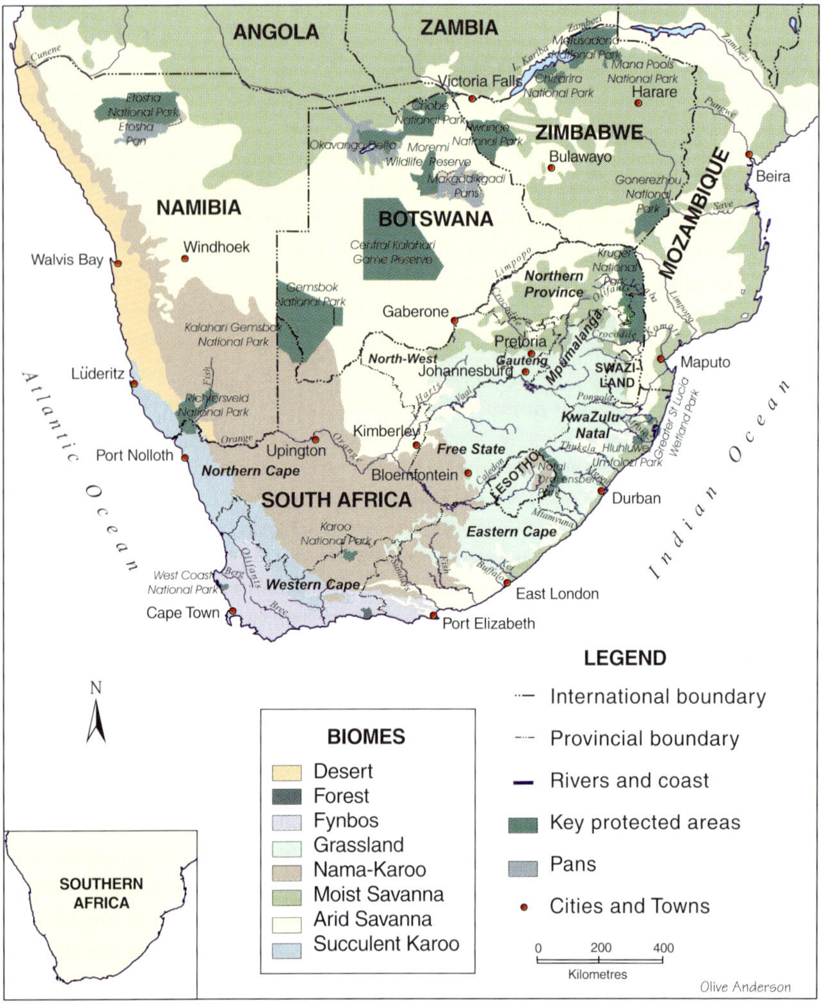

ANGOLA
ZAMBIA
ZIMBABWE
NAMIBIA
BOTSWANA
MOZAMBIQUE
SOUTH AFRICA
SWAZI-LAND
LESOTHO
Atlantic Ocean
Indian Ocean
Victoria Falls
Harare
Bulawayo
Beira
Windhoek
Walvis Bay
Lüderitz
Port Nolloth
Gaberone
Pretoria
Johannesburg
Maputo
Kimberley
Upington
Bloemfontein
Durban
East London
Port Elizabeth
Cape Town
Northern Province
North-West
Gauteng
Mpumalanga
KwaZulu-Natal
Free State
Northern Cape
Eastern Cape
Western Cape
Etosha National Park
Etosha Pan
Chobe National Park
Okavango Delta
Moremi Wildlife Reserve
Makgadikgadi Pans
Hwange National Park
Chizarira National Park
Mana Pools National Park
Gonarezhou National Park
Central Kalahari Game Reserve
Gemsbok National Park
Kalahari Gemsbok National Park
Richtersveld National Park
Kruger National Park
Greater St Lucia Wetland Park
Hluhluwe-Umfolozi Park
Natal Drakensberg Park
Karoo National Park
West Coast National Park
Cunene
Zambezi
L. Kariba
Limpopo
Crocodile
Olifants
Orange
Fish
Vaal
Harts
Caledon
Thukela
Mfolozi
Mkomazi
Mzimvubu
Kei
Buffalo
Sundays
Berg
Breë
Save
Pungwe
SOUTHERN AFRICA
N
BIOMES
Desert
Forest
Fynbos
Grassland
Nama-Karoo
Moist Savanna
Arid Savanna
Succulent Karoo
LEGEND
International boundary
Provincial boundary
Rivers and coast
Key protected areas
Pans
Cities and Towns
0
200
400
Kilometres
Olive Anderson

Section One

Introduction to bats

Chapter One

Why bats?

Bats are the second most diverse order of mammals worldwide, with nearly 1 000 described species. Only the rodent order contains more species. Yet bats remain poorly studied and misunderstood in most parts of the world, including the African continent. Despite their fundamental ecological and economic value, declines in bat populations have been documented in certain parts of the world, often resulting from human persecution (see Chapter Seven). For the most part, however, data on the conservation status of bats are simply lacking.

It is only through educating the wider public about the real benefits, beauty and diversity of bats, by dispelling harmful myths, and by recruiting enthusiastic amateur bat-watchers to assist with research, that we can change the situation regarding bats. This approach has succeeded in some European countries, North America and Australia, where numerous public bat conservation groups have placed bats firmly in the public eye, and at the forefront of national conservation efforts. Through the 1990s, three South African bat interest groups have been formed, and these have already demonstrated that educated amateur bat enthusiasts can play a critical role in enhancing public awareness, as well as boosting research and conservation efforts in southern Africa (see Chapter Seven).

This book was written to support this encouraging growth of bat conservation groups, to raise public appreciation of bats in southern Africa, and to provide amateur and professional bat-workers with both a species identification guide and a handbook on bat-watching.

Thus, whilst also attempting to cater for the more serious bat biologists (species accounts summarise information on reproduction, feeding, size measurements, and the echolocation calls of southern African species), this book was primarily written for the budding bat-watcher. Just like birds, bats are all around us and can easily be appreciated and studied without harming them, particularly with the aid of bat detectors (see Chapter Eight). New members of the Durban Bat Interest Group (DBIG), as well as those who take part in our public outings, are often surprised and delighted to discover that Wahlberg's epauletted fruit bats, banana bats, and Mauritian tomb bats can be viewed in broad daylight within a kilometre of Durban's city centre. Many of our members have already been 'hooked' by bat detectors!

The first section of this book aims to dispel some prevalent myths about bats (see Chapter Two) and to introduce the reader to some of the remarkable adaptations which have enabled bats to conquer their nocturnal niche (see Chapter Three). Those of us who have researched and cared for bats have found them to be clean, endearing, intelligent, caring mothers; in fact nothing like most popular portrayals. Their astonishing adaptations never cease to amaze and delight us. Just take their intricate navigation system, echolocation, many times more effective than man's most sophisticated radar system, or the ability of female bats to store viable sperm in their oviducts or uterus for months so as to coincide the birth of their young with spring's abundance of insect food. Witness their incredible powers of hibernation, their agile flight and their unbelievable longevity (up to 34 years or more).

The second section provides an illustrated identification key to the families, genera and species of bats (see Chapter Five), as well as species accounts for all 74 extant species of southern African bats (see Chapter Six). Both colour, as well as black and white illustrations, tables of measurements, distribution maps, and sonograms (time-frequency graphs of echolocation calls) are included to assist with identification.

The third section of the book includes a brief overview of bat conservation globally and locally, and is also designed to provide practical guidance for those eager to get started with bat-watching and 'hands-on' conservation projects such as bat groups, bat houses, bat detectors, and dealing with householders' bat nuisance problems. There are many ways to get started. Join or start a bat group (see Chapter Seven), report bat roosts known to you to a bat group or natural history museum, buy an inexpensive bat detector (see Chapter Eight) and find your neighbourhood come alive with bat ultrasound at night, assist your neighbours in identifying bats in their roofs and persuade them to keep them (see Chapter Nine), or buy or build yourself a few bat houses (see Chapter Ten). Further references and contacts are provided on where more detailed technical information can be obtained. The section on bat detectors and echolocation is particularly important, as bat detectors are being used increasingly worldwide to identify species of bats 'on the wing'. As we develop our growing library of calls for southern African species, bat detectors will become indispensable aids in identifying species.

For too long we have ignored or feared bats. In our prejudices and ignorance, and through our unprecedented population growth and habitat destruction, we are the cause of their decline. But we can also be their salvation. The battle for their survival will be won or lost in the hearts and minds of the average citizen. Enthusiastic bat friends worldwide, and in southern Africa, are coming to the rescue of these fragile nocturnal mammals. Britain boasts 90 bat clubs with 2 000 members. In southern Africa, we are just beginning. You can make a difference. Through becoming an educated bat-watcher, I personally invite you to make a difference.

Chapter Two

Bat myth and reality

Here is a quick test. Answer true or false:

1. Bats are blind?
2. Bats get tangled in human hair?
3. Blood-sucking vampire bats are found throughout the world?
4. Bats are messengers of evil, and associated with witchcraft?
5. Bats attack humans?
6. Bats nibble your ears at night?
7. Most bats carry rabies?
8. You can get a fatal lung disease from bat guano in caves or roof spaces?
9. Bats and their droppings carry parasites which cause rashes and disease in humans?

If you answered true to any of the above (they are all myths), you are a victim of 'bat phobia'. Bat phobia is an unfortunate state of mind resulting from generations of ignorance and misinformation, stemming largely from writers such as Bram Stoker, author of the infamous book, *Dracula*. However, there is a cure for bat phobia: read the rest of this book.

Sadly, apparently respectable people of every race, sex, or age have been found guilty of believing every one of the myths listed above. Let us re-visit these myths, and debunk myth with scientific fact.

1. Bats are blind?

Even 'educated' South Africans continue to believe this myth. A couple of years ago, I was addressing a grade five class during a field outing. In response to my first question: 'Are bats blind?', several students shook their heads but their teacher blurted out confidently: 'Of course they're blind, they use radar.'

Although insect-eating bats can use echolocation (a system akin to sonar) to orient themselves at night when lighting is unpredictable, none of the over 900 species is blind. All bats have eyes and many see very well, indeed as well as humans or owls.

2. Bats get tangled in human hair?

Many South Africans, especially women, continue to believe this. One brain-washed woman recounted how all the girls at her boarding school were instructed by their teachers to wear bonnets over their hair to ward off bats when crossing the courtyard at night. A bat-friendly couple from Westville (Durban) described the reaction of female visitors when shown a pair of fruit bats roosting under their thatched gazebo; they invariably clutched their heads protectively.

If you've ever tried to catch a flying bat with a hand net in a cave filled with thousands of bats, as I have, you'll soon discover how sophisticated their navigation system is. Although I could feel the air from their wings passing close by, not one touched me, nor was I even able to catch a single bat in my net.

3. Blood-sucking vampire bats are found throughout the world?

In the world there are three species of vampire bats. They feed on blood and occur only in both South and Central Americas. They do not occur in Africa, Asia, Australia, North America, or Europe, not even in Transylvania!

Yet many people insist on believing that vampire bats occur worldwide. On one occasion, I was called out to a Durban house to identify a 'vampire bat' which had been caught by a little boy. The boy's mother would hear nothing of my insistence that vampire bats do not occur in Africa. When I arrived, there was much distress. The 'vampire bat' had escaped in the boy's bedroom. A thorough search proved fruitless, and when I left, the mother was more convinced than ever of the missing bat's identity.

A local newspaper featured a picture of a Durban Bat Interest Group member holding a yellow house bat. The member made a point of telling the reporter that vampires were *not* found in South Africa. Big mistake. The caption proudly announced: 'Kim Mari holding a South African vampire bat'.

Recently, medical research has resulted in the isolation of an anti-coagulant factor in the saliva of the common vampire which is being used to help thin the blood of heart-attack victims. The product name is, unsurprisingly, 'Draculin'.

4. Bats are messengers of evil, and associated with witchcraft?

An association between bats and witchcraft is quite prevalent among traditional cultures in southern Africa. The cleaner at work was once horrified to learn, whilst cleaning my office, that a live bat was being kept in the office. She was worried that one of her friends would see her in the company of a bat and thereafter brand her a witch. Ground up remains of bats have been used by herbalists as powerful 'muti', supposedly capable of inducing madness if added to the drink of another. Certain satanic groups in the Durban and Pietermaritzburg areas have a penchant for using bat caves for their rituals. While recently watching a documentary on ancient Caribbean cultures, I noted that bats were regarded as messengers from the dead. The tribe's religious shamans undertook their psychic consultations in the dark recesses of bat caves.

On the bright side, it is worth noting that in China, Cuba and in the Pacific Islands, bats are regarded as good luck charms. The Chinese word *fu* for a bat also means good fortune. Bat symbols figure prominently in Chinese designs on fabrics, dagger handles, pottery and other crafts. Apparently, even the bat symbol on the logo of Bacardi Rum was initially a symbol of good luck instituted by the original Cuban family owners of the business. Is it not time we took a lesson? Bats are no more evil than a pet pekingese.

5. Bats attack humans?

Like any cornered wild animal, bats may bite in self-defence if picked up. However, bats are naturally gentle and non-aggressive animals and unprovoked attacks by bats on humans are unheard of. Nevertheless, in many cultures, bats are perceived to be aggressive. I will never forget Mrs Pillay from Durban, whose family were kept imprisoned in their home after dark by bats living in their roof space, which apparently swooped down to attack them with every intent of burrowing under the skin, whenever they ventured outdoors at night. Apparently the entire community believed this myth, although no one had ever actually sustained or witnessed an attack.

6. Bats nibble your ears at night?

I have repeatedly heard this myth among superstitious rural South Africans living in KwaZulu-Natal. Some people even wear balaclavas while sleeping to prevent having their ears nibbled. I once asked a class of schoolchildren if any of them, their parents, or friends had ever known anyone to be bitten. Just guess the answer?

7. Most bats carry rabies?

Rabies is a fatal disease which is spread by the bite of an infected animal. The bite must break the skin for the virus to enter the system, but even saliva on existing cuts and mucous membranes can be a risk. The incubation period is normally 20–90 days, and death usually results after about 14 days after incubation is complete.

Scientists have identified seven strains, or serotypes, of rabies and 'rabies-like' viruses (called lyssaviruses), certain of which may, very rarely, be harboured by bats:

Serotype 1: typical rabies
Serotype 2: Lagos virus
Serotype 3: Mokola virus
Serotype 4: Duvenhage virus
Serotype 5: European bat lyssavirus type 1
Serotype 6: European bat lyssavirus type 2
Serotype 7: Australian bat lyssavirus

In southern Africa, 'typical' rabies (serotype 1) is carried by dogs and wild mammals, but has never been detected in bats. Two lyssaviruses, Lagos virus and Duvenhage virus, have been isolated from bats in South Africa (although it is not known whether bats are asymptomatic carriers, or whether they themselves die from the virus). No cases of Lagos bat virus in human victims have been reported. Nevertheless it is almost certain to be fatal if transmitted. A single human death from encephalitis resulted from Duvenhage virus: a certain Mr Duvenhage who was apparently bitten by a sick, grounded insectivorous bat at Warmbaths in the 1970s. A recent study of 580 South African bats found that not one tested positive for a rabies-like virus.

In the past two years, two human deaths in Australia have resulted from the newly discovered Australian bat lyssavirus. In both cases, the victims were persons involved in the rehabilitation of flying foxes. Scientists believe this lyssavirus has probably been circulating in that bat population at a very low level for thousands of years.

According to Dr Merlin D. Tuttle of Bat Conservation International: 'In the period from 1980 through 1999, public health records indicate that only 22 people in the United States and Canada contracted bat strains of rabies. In contrast, approximately 400 people died in the same period, in the United States alone, just from the attacks of healthy dogs, not to mention 200 000 from spouse attacks.'

Safety guidelines

1. Don't handle bats (or any wild mammals) unless you have to;
2. Wear gloves, or handle bats with a towel, to prevent getting bitten;
3. Get innoculated with the rabies pre-exposure vaccine if you are going to be handling any wild mammals;
4. Soap and iodine may inactivate the rabies virus, use them liberally on any bites or scratches;
5. Be prepared to have any mammal that bites a person put down and tested. Contact the Onderstepoort Veterinary Institute near Pretoria, or your nearest government or municipal health department.

    Address:
    Rabies Laboratory
    Private Bag X5
    Onderstepoort 0110
    South Africa
    Tel: (012) 5299440
    Fax: (012) 5299390

8. You can get a fatal lung disease from bat guano in caves or roof spaces?

I have often heard, from the public, horrific stories of this allegedly fatal lung disease which will inflict anyone who enters bat caves.

In reality, acute benign histoplasmosis (the correct medical term) is a mild lung disease which can be caused by inhaling spores of the fungus *Histoplasma capsulatum*, living in moist bird or bat guano. It is quite common among pigeon fanciers as well as cavers and bat scientists who have had exposure to large quantities of dust from bat guano in caves. According to South Africa's medical expert on histoplasmosis, Dr Steven Craven of Cape Town, it is almost never fatal, and is only potentially life threatening in one out of some 100 000 cases. Nevertheless, as a precaution, anyone entering caves or roof spaces where a large accumulation of moist guano is present should wear a protective pathogen filter mask capable of filtering particles as small as 1 micron diameter. Anyone experiencing flu-like symptoms a couple of weeks after unprotected exposure to a large quantity of dust from moist bat or bird guano, should contact a medical doctor for advice and treatment. Those with deficient immune systems (e.g. those with HIV infection) are at a much higher risk of medical complications related to histoplasmosis.

9. Bats and their droppings carry parasites which cause rashes and disease in humans?

A common, albeit irrational, fear of bats relates to the imagined health risks (disease or allergies) attributed to the parasites of bats. One medical doctor swore that bats in his home ceiling were causing an itching allergy in his family due to 'fleas' on the bats and in the bat guano. He was concerned about possible outbreaks of disease. After bats and their guano from his ceiling were tested and shown to be remarkably free of parasites, he sheepishly admitted that the problem had gone away.

Bats harbour some fascinating but completely harmless parasites including scores of species of strange wingless flies that, under a microscope, look like aliens from a Steven Spielberg movie. These flies (of the families Strebilidae and Nycteriibidae) are specific to bats and occur on no other species, including man. One look at their weak mouthparts is enough to convince anyone they are incapable of biting through human skin. Bats do not harbour lice, but may harbour the occasional bug (akin to bed bugs) or flea which rarely bite any animal other than another bat. Usually only one or two parasites at most occur on a single bat, and diseases, rashes or any other conditions in humans known to be associated with bat parasites are extremely rare indeed, and very unlikely to be transmitted to someone who is not in close continuous physical contact with large numbers of bats; this includes almost everyone. When bats move away from a roost, their parasites either follow them or soon die.

Chapter Three

Bat biology

Being small, night-flying, warm-blooded mammals places huge energy demands on bats, particularly in temperate regions where cold weather and low insect numbers force bats to hibernate in winter or migrate prior to winter. Studies have shown that at least 70% of some species of bats may die in their first year. Yet bats are equipped with incredible anatomical and physiological adaptations for survival: wing and body structure, hibernation, echolocation and sperm storage, to mention but a few.

This chapter touches on some of the important features of bats which have allowed them not only to survive, but to become the second most diverse group of mammals (next to rodents), containing almost 1 000 species worldwide. Bats are true marvels of nature, and an appreciation of this goes a long way towards engendering a more sympathetic public response to bats.

ORIGINS AND CLASSIFICATION

Fifty million years ago, during the Eocene, there were at least 13 species and 13 families of bats. Eocene fossil bats are known from North America, Europe, and Australia. By then bats were already fully developed as bats.

Bats are classified as mammals. Although Linnaeus, the father of taxonomy, classified bats as primates (because they have a pair of nipples born on the chest), bats are now classified in a unique order of mammals, Chiroptera (Greek for 'hand-wing'). Two sub-orders, or major groups, of bats, are recognised: the larger Old World fruit-eating bats, Megachiroptera, commonly called megabats; and the smaller, predominantly insect-eating Microchiroptera, commonly called microbats. Megabats have claws on the thumb and forefinger, large eyes, a reduced or absent tail, and a dog-like muzzle, whereas microbats have only a thumb claw, smaller eyes, and a tail which may or may not be enclosed in the wing membrane.

A recent scientific hypothesis, hotly contested by some bat scientists, suggested the possibility that megabats evolved from primates, while microbats evolved separately from a shrew-like, tree-living ancestor. This would mean that flight has evolved twice in two separate lineages of bats. At the time of writing, this debate continues.

DIVERSITY AND DISTRIBUTION

Bats vary considerably in size and appearance, and in the anatomical details of their face, ears, teeth, tail, and wings. The world's smallest bat, the 2 g hog-nosed bat from Thailand, is 600 times smaller than the largest species, the giant flying fox which weighs up to 1.2 kg, with a wingspan of 1.8 m. Some bats sport an impressive variety of complicated facial structures termed noseleafs. In general bats' ears are greatly enlarged in proportion to their overall body size when compared for example with humans. This is undoubtedly related to their role as efficient sound receivers during echolocation. However, ear length varies from relatively short in long-fingered bats to long and rabbit-like in shape in the common slit-faced bat. The tail may be absent or present, short to long, and enclosed to varying degrees in a thin tail membrane. While one species of bats is hairless, hair length and colour can also vary among bats. Most bats have fairly short grey, brown or black hair, but bats like the woolly bats and hairy bats of Africa have long, soft, upstanding fur; in the case of the hairy bats this is often an attractive coppery red colour. The yellow-winged bat from Central and East Africa has striking yellow wings. Species such as the butterfly bat and Welwitsch's hairy bat of southern Africa have distinctively patterned wings.

Bats make up a quarter of all mammal species and are found throughout the world except for extreme polar and desert habitats. Species diversity of bats is concentrated in equatorial regions with progressively fewer species encountered with increasing latitudes in temperate climates. Southern Africa boasts 74 species, of which 56 have been recorded in South Africa. In the Durban metropolitan region alone, 21 species of bats have been recorded, more than the total list of 16 bat species found in Britain.

Although bats occur throughout southern Africa, they vary from region to region in the numbers of species (species richness) found in different vegetation types, or biomes.

Based on collections of bats in museums, C. Gelderblom and co-workers plotted differences in species richness of bats across South Africa. Most of the bat 'hotspots' (quarter-degree grid squares with a high number of species) occurred in the savanna regions of northern KwaZulu-Natal, the lowveld of Mpumalanga, and the Northern Province. The table following provides approximate totals for the number of species found in each of the major biomes making up South Africa, from which it is clear that the savanna regions have the most species. The table also lists species numbers for each of the countries within southern Africa, as well as some of the better surveyed provinces of South Africa. Countries such as Mozambique remain very poorly sampled and further collecting will probably lead to a large increase in the species total. Notwithstanding sampling biases, a pattern of increasing species richness from south to north (i.e. as one approaches the equator) is evident. For example, Zimbabwe, although much smaller in area than South Africa, has more species of bats. Nevertheless, two important South African endemic species (found only in South Africa) are known from the fynbos and Karoo biomes of South Africa, the Cape horseshoe bat and Lesueur's hairy bat respectively.

Biome or province (South Africa), or country	Total number of species	Number of species exclusive to region or biome[1]
Fynbos	18	1
Nama-Karoo	17	1
Succulent Karoo	12	0
Grassland	31	0
Savanna	51	15
Former Cape Province, South Africa[2]	29	2
KwaZulu-Natal, South Africa	38	2
Free State, South Africa	10	0
Former Transvaal[3], South Africa	39	0
South Africa (combined)	56[4]	5
Lesotho	8	0
Swaziland	20	0
Zimbabwe	61	6
Botswana	34	1
Namibia	26	3
Mozambique	45	1

[1] 'Exclusive' taken to mean 90% or more of southern African records fall within the defined area
[2] Comprises the Western Cape, the Eastern Cape, the Northern Cape, and parts of the North-West
[3] Comprises Gauteng, the Northern Province, Mpumalanga, and part of the North-West
[4] Including two species known only from very old type-specimens

ECHOLOCATION

In Africa, all insect-eating bats, and one species of cave-dwelling fruit bat, the Egyptian fruit bat, communicate and navigate using echoes bounced back from high frequency, ultrasonic clicks and squeaks emitted through the mouth or nostrils – a system known as echolocation. The ultrasonic calls emitted by bats range in frequency ('pitch') from 20 to 210 kilohertz (one kilohertz, or 1 kHz, equals 1 000 vibrations per second) and are generally inaudible to humans (who hear sounds of up to 20 kHz). Even though we cannot hear them, bats' echolocation calls vary considerably in amplitude, with 'whispering bats' having soft calls and bats like horseshoe bats having very loud calls, with amplitudes comparable to the sound of a jackhammer at close range. The echolocation system of bats is vastly more efficient than any man-made sonar system. The United States navy is conducting ongoing research into the echolocation system of bats so as to model its underwater sonar systems more closely to that of bats.

Today we can use bat detectors to convert the higher pitched bat sounds into signals that we can hear. This provides us with a window to the fascinating lives of bats (see Chapter Eight).

MIGRATION

In temperate regions (and some subtropical regions) cold winters force bats to migrate or hibernate. Some bats migrate, many making local, short (<50 km) or medium distance (<500 km) movements between winter and summer roosts. In southern Africa, the seasonal appearances and disappearances of straw-coloured fruit bats are examples of migrations which appear to occur as bats follow their food supplies.

In Gauteng, the Northern Province, and the Western Cape of South Africa, Schreibers's long-fingered bats migrate up to 260 km. The movements occur between warmer maternity caves where females give birth in summer (e.g. the De Hoop Guano Cave in the Western Cape, and several caves in the lowveld of the Northern Province), and colder caves in winter, where mating and hibernation occurs (e.g. several caves in the interior of the Western Cape, and on the highveld of Gauteng). Temminck's hairy bat undertakes similar seasonal migrations.

Egyptian fruit bats may also make migrations of hundreds of kilometres. An individual tattooed at caves in the Tzaneen area of Mpumalanga was subsequently recovered at Mtunzini on the northern coast of KwaZulu-Natal. Population numbers of the Mpumalanga colonies decrease during winter and increase again in summer, while, conversely, a known colony at Mission Rocks Bat Cave at Lake St Lucia on the northern coast of KwaZulu-Natal decreases in size dramatically (from >5 000 to <300) at the beginning of spring, and increases again during winter. It seems highly likely that the same bats migrate from winter roosts on the warm KwaZulu-Natal coast to summer roosts in the Mpumalanga lowveld.

TORPOR AND HIBERNATION

Although bats are warm-blooded mammals, and can maintain a constant body temperature, some show a remarkable degree of tolerance for fluctuation in body temperature (thermolability). This may happen on a daily or seasonal basis. Torpor is the ability of bats and other mammals to drop their body temperature in response to cooling and/or reduced food availability. Torpid bats have body temperatures colder than normal and close to environmental (ambient) temperature. Daily torpor is common in plain-nosed (family Vespertilionidae), horseshoe (family Rhinolophidae) and leaf-nosed (family Hipposideridae) bats, and in some free-tailed (family Molossidae) bats, as is seasonal hibernation, which involves prolonged periods of torpor. The other bats from southern Africa are more conventional thermoregulators and cannot enter torpor.

Recent research by G. Bronner and co-workers has shown that Angolan free-tailed bats can tolerate extremely hot temperatures under tin roofs in the South African lowveld by allowing their body temperature to rise during the day in order to save energy required for cooling (the reverse of torpor).

Although a few bats migrate to warmer climates for winter, as do some birds, many bats actually move to colder climates for winter. The answer to this apparent riddle is bats' ability to hibernate, that is, to shut down their body's metabolism to a basic level in response to winter cooling, thus allowing their body temperature to fall to that of their environment, or to just above freezing in sub-zero temperatures. The colder the outside (ambient) temperature, the greater the saving of energy effected by bats through the lowering of their internal temperature. By switching off their energy-expensive heating system during hibernation, bats can survive cold winters when insects (i.e. fuel reserves) are very scarce. Prior to hibernating, in late summer, bats will eat voraciously to build up their fat reserves, which may comprise up to 28% of their total body weight just prior to hibernation, and which may have to tide them over for up to six months during hibernation. Disturbing hibernating bats can have tragic consequences. Bats budget their fat reserves very precisely and forced arousal from hibernation consumes huge amounts of energy reserves, and can result in bats starving before the end of winter. From studies in the northern hemisphere, one arousal costs the energy that could cover 60 days of hibernation.

REPRODUCTION

Reproductive patterns of bats show some unique features compared with mammals generally, linked to the demands of hibernation which necessitate interruptions of the reproductive process.

Worldwide, average length of bat gestation (the period from fertilisation to birth) varies from 44 days in the common European pipistrelle to seven months in vampire bats, but typically it is around two months. Gestation of southern African bats varies from 60 days in the little free-tailed bat to an effective seven to eight months in species which show delays in the development of the embryo.

In hibernating bats, mating occurs in autumn, just prior to the winter hibernation period, leading to an interruption of the normal reproductive processes of pregnancy. Three patterns of delayed reproduction can be recognised in hibernating species:

1) **Sperm storage**. After mating occurs in autumn, viable spermatozoa are stored and nourished in the female's uterus or oviducts throughout the hibernation period. Eggs are only fertilised by the spermatozoa once the females have ovulated after arousal from hibernation in spring. Most temperate-climate bats use this phenomenon, as do many southern African plain-faced and horseshoe bats. The burden of sperm storage can also be shared between the sexes. In banana bats, rusty bats, Temminck's hairy bat and most rhinolophids, viable sperm is stored predominantly in the female reproductive tract, while in Cape serotine and Schlieffen's bat, sperm is stored in both the male and female tracts. At the other extreme, only male Cape horseshoe bats store sperm, and mating, ovulation, and fertilisation occur only after winter.
2) **Delayed implantation**. Mating, ovulation, and fertilisation occur in autumn, but the implantation and subsequent development of the resulting blastocyst (fertilised ovum) is delayed until after hibernation is complete. In southern Africa, Schreibers's long-fingered bat and the lesser long-fingered bat exhibit this phenomenon.
3) **Retarded embryonic development**. Normal mammalian fertilisation and implantation occur, but the embryo's growth is retarded during hibernation and only resumes after hibernation is over. In southern Africa, Sundevall's leaf-nosed bats exhibit this phenomenon.

Among non-hibernating bats, two basic patterns of 'normal' mammalian reproduction are common in southern African bats:

1) **Seasonal polyoestry**. More than one pregnancy and birth season occurs during the restricted breeding season (e.g. Wahlberg's epauletted fruit bat, Angolan free-tailed bat and little free-tailed bat).

2) **Seasonal monoestry**. Only a single pregnancy and birth season occurs during the restricted summer breeding season (e.g. the majority of bats, including most plain-faced bats).

Whether they are hibernators or not, and whether they are monoestrus or polyoestrus, most southern African bats seem to have a single young per litter (see Chapter Six). Interestingly, while all polyoestrus species give birth to a single young per litter, some plain-faced bats with a monoestrus pattern tend to have more than one young at a time. Banana bats, rusty bats, yellow house bats, Schlieffen's bats and Cape serotine bats commonly give birth to twins or triplets, and even sometimes quadruplets in the case of Cape serotines.

Species with widespread distributions may have multiple births in equatorial parts of their range in Africa, but only one or two birth periods in more temperate regions; this seems to be the case in the Mauritian tomb bat.

In the gregarious species, pregnant females typically form nursery (maternity) colonies in summer. Here, the young are born, often with the mother clinging upside down to the substrate with toe claws and one wing claw to maintain a semi-horizontal position. The wing and tail membranes are used to catch the new-born. On average, a new-born bat weighs 22% of its mother's weight (compared with 8% in other mammals), and has adult-sized hind feet as well as deciduous teeth to assist with clinging onto the fur on the underside of the mother. The young may be carried for one or two weeks while the mothers forage, but thereafter they are left in nursery clusters. Research at Bracken Caves in Texas has shown that mothers recognise their own young, though there may be some 5 000 pups per square metre.

Courtship among bats is sometimes more subtle than, say, birds, but occurs nevertheless. Pipistrelles in Europe have been shown to engage in melodious ultrasonic courtship 'songs' to attract females. Male epauletted bats sing from well-situated calling stations and flash large fluffs of white shoulder fur to attract mates; they probably also use 'deodorant' secretions from glands contained in their shoulder pouches.

In some species, it seems that favoured mating 'turfs' guarded by certain males are more attractive to females than the males themselves. In Britain, individual greater horseshoe bats have been seen to guard the same spot at cave entrances over a large number of years.

Mating systems of bats are quite variable. A common pattern is polygyny, in which one dominant male mates with a harem of females. This pattern is known in South American common vampire bats (*Desmodus rotundus*), but may also be common in southern Africa. It appears to be the case, for example, in the roof-roosting, large-eared free-tailed bat (*Otomops martiensseni*), where a dominant larger-sized male usually confronts approaching intruders with a piercing squeak. Other members of the colony are typically females with juvenile or sub-adult males and females. In a large Kenyan cave colony of sheath-tailed bats (*Coleura afra*) exceeding 50 000 individuals, the bats were arranged in clusters, the majority of which were harems attended by a single adult male, sometimes with a second 'satellite male' on the periphery. Monogamy, where male and female partners form a life-long mating bond, is rare in bats and has not been shown to occur in any southern African species. Some species of bats are promiscuous, with matings occurring more or less randomly in the population.

Bats are, for their size, the slowest reproducing mammals on earth. On average, mother bats rear only one young per year, and some are two or three years old before they have their first young. For their size, bats are also incredibly long-lived. Lifespans of 34 years have been recorded, although the normal is some 15–20 years. Mortality of bats is highest in their first year, when some 70% may die.

PREDATORS

Before discussing what bats eat, an interesting question often asked is what eats bats. Bats' natural mortality is high in their first year, although remarkably low in subsequent years. Animals that will hunt bats include snakes, owls, other birds of prey, crows, genets, domestic cats and humans. Some bats, for example the giant slit-faced bat of central and southern Africa, eat other bats. The bat hawk is a specialised predator of bats.

DIET AND FEEDING BEHAVIOUR

In southern Africa, Microchiroptera feed on a wide range of insect prey (with the exception of the giant slit-faced bat which takes vertebrate as well as invertebrate prey), while all Megachiroptera are frugivorous (fruit-eating), nectivorous (nectar-feeding), or eat leaves and flowers. A brief introduction to diet and feeding behaviour of bats follows, but information for individual species is discussed in more detail in Chapter 6.

Frugivorous and nectivorous bats

Megachiropterans feed on the fruits, leaves, flowers, or nectar of a wide range of indigenous trees, often showing preference for the fruits of figs (*Ficus* spp.). The fruits of yellow-wood trees (*Podocarpus* spp.) are also favoured. Soft-fleshed cultivated fruits are highly favoured, including mangos, guavas, pawpaws, avocadas, litchis, bananas, dates of palms, and even the berries of the *Syringa* tree. Bats prefer ripe fruits, and, since most commercial fruits are picked under-ripe for shipping, fruit bats generally should pose no significant risk to fruit orchards. However, there are apparently exceptions to this generalisation, for example, Egyptian fruit bats are said to cause damage to litchi orchards in South Africa. In residential areas, fruit bats can be a nuisance by defaecating on the walls of houses; since fruit bats have excellent vision, illuminating the walls has proven to be an effective deterrent.

Carnivorous and blood-feeding bats

A few bats, including the giant slit-faced bat of Africa, are carnivorous. The giant slit-faced bat feeds on smaller vertebrates such as fish, frogs, mice, birds, and even other bats. Elsewhere in the world, frog-eating bats are known to respond to the mating calls of frogs, while fish-eating bats have a specialised echolocation system to detect fine ripples on the water surface caused by fish, and sharp, hooked hind claws which they can use as a gaffe to snatch their prey. The Australian false vampire bat catches and eats small rodents and birds. As mentioned earlier, blood-feeding vampire bats are restricted in their distribution to Central and South Americas.

Insectivorous bats

Some 70% of bat species worldwide eat insects. Depending to varying extents on their body size, jaw construction, wing shape, foraging behaviour, habitat use, or the nature of their echolocation calls, different insectivorous species feed on different groups (orders) of insects. Studying the diet of insectivorous bats requires detailed microscopic examination of faecal remains or stomach contents. Morphological measurements of the skull and mandible can also enable researchers to predict the diet. Strong robust jaws and long canines are indicative of hard-shelled prey, while a weak jaw, with poorly developed coronoid process, is indicative of soft-bodied prey such as moths. In general smaller bats are limited to smaller prey while larger bats can take a wide size range of prey.

Aerial feeders, such as most plain-faced bats (Vespertilionidae), sheath-tailed bats (Emballonuridae) and free-tailed bats (Molossidae), hunt flying insects exclusively on the wing, while gleaners snatch their stationary prey from branches or the ground. The last-mentioned are capable of slower, more manoeuvrable flight in confined spaces. Some species, such as the common slit-faced bat, and possibly Sundevall's leaf-nosed bat and certain horseshoe bats, are capable of both aerial feeding and gleaning. Some of these slow-flying and manoeuvrable species, for example many horseshoe bats, also 'perch-hunt' by hanging on a perch and making lightning quick attacks when prey is detected moving past.

H.D.J.N. Aldridge and I.L. Rautenbach defined four different classes of aerial feeders depending on their degree of manoeuvrability, i.e. their ability to fly in confined or cluttered spaces without stalling or colliding with obstacles: 'open', 'woodland edge', 'intermediate', and 'clutter' foragers. These groups differ in their echolocation abilities as described in more detail in Chapter Eight.

1) Open foragers, such as free-tailed bats and tomb bats, fly rapidly and feed in open spaces above the vegetation canopy, using low frequency, long duration, narrow-band echolocation calls to detect larger prey at long range.
2) Woodland edge foragers, like many of the smaller plain-faced bats, typically forage at the edges of the vegetation canopy, where their short, broad-band echolocation calls equip them for slower, more manoeuvrable flight and more accurate, short-range target detection than the first group.
3) Intermediate foragers, like the yellow house bat, lesser yellow house bat and the butterfly bat, can forage in both open spaces and more cluttered situations within the vegetation canopy. Their echolocation calls possess both narrow-band and broad-band components.
4) Clutter foragers have broad, short wings, and high echolocation frequencies to allow them to fly slowly and manoeuvre in dense clutter. Clutter foragers include short-eared trident bats, Sundevall's, and Commerson's leaf-nosed bats, as well as all horseshoe bat species.

Both aerial clutter foragers and gleaners tend to use caves as day-time roosts and establish temporary feeding stations, termed night roosts, during their foraging period.

Clutter foragers typically use constant frequency (CF) echolocation calls which exploit the Doppler shift phenomenon to act as 'motion detector systems', allowing them to detect the fluttering wings of moths and other flying insects against a dense vegetation background (see Chapter Eight).

Moths feature prominently in the diet of many bats. Tympanate moths, those possessing hearing organs termed tympani, have evolved the ability to detect the echolocation calls of bats and take avoidance measures to prevent being caught. In a sort of 'evolutionary arms race', some bats which are moth specialists (e.g. the short-eared trident bat, the tomb bat, and the large-eared free-tailed bat), have evolved specialised echolocation calls to avoid detection by moths (see Chapter Six).

ROOSTING HABITS

Wherever you live, even in big cities, bat roosts can be located, especially once you become familiar with the individual roosting habits of different species. The Durban metropolitan region currently harbours at least 19 species of bats, some of which can be spotted within a kilometre or two of the city centre. On one morning outing, we visited Mauritian tomb bats roosting in royal palm trees and Chinese fan palm trees at the Botanic Gardens, two colonies of Wahlberg's epauletted fruit bats near a major shopping centre, one in royal palm trees, and one (of at least 90 bats) under the eaves of a double-storey house, and a banana bat colony roosting in a grove of bananas in the middle of the city's main racecourse. On another evening outing, we listened with bat detectors while three species of free-tailed bats, including the rare large-eared free-tailed bat, emerged from a double-storey apartment building in a Durban suburb.

The day-roosting habitats of bats can be classified as: foliage; hollows; crevices; and specialised roosts. Some tips are given below as to where and how to spot bats in each of these roosting situations. If you know of bat roosts in your area, your nearest bat interest group or bat scientist would be grateful to know the details (see Important contacts).

Foliage-roosting bats

Most Megachiroptera, as well as tomb bats, banana bats, butterfly bats, and Welwitsch's hairy bats hang up or cling onto surfaces in trees or shrubs. While fruit bats, butterfly bats, and Welwitsch's hairy bats hang by their hind claws from the undersurface of leaves or branches (the last two hanging in a disguised manner among clumps of leaves). Mauritian tomb bats roost face-down, anchored by their hind claws but with their bellies, thumb claws, and hind claws in contact with the surface of a tree or wall; and banana bats cling with their ventral surface in contact with the smooth surface of unfurling banana leaves where they roost (Figure 1). More details of the roosting behaviour of these species are given in Chapter Six.

Figure 1 Potential roosting sites of bats

1. Banana bat
2. Epauletted fruit bat
3. Free-tailed bat, yellow house bat
4. Mauritian tomb bat
5. Lesser yellow house bat, serotine bat
6. Free-tailed bat, serotine bat
7. Slit-faced bat, horseshoe bat
8. Woolly bat
9. Welwitsch's hairy bat

Hollow-roosting bats

This category includes bats that occupy underground caves, hollows in trees, and man-made hollows such as roofs and basements of houses, tunnels or cavities in dam walls, and abandoned mine shafts. Some plain-faced bats, sheath-tailed bats, and free-tailed bats roost in tree hollows. Radio-tracked lesser yellow house bats were found to occupy hollow mopane trees in the Kruger National Park. While a large amount of research is being carried out in the United States and Canada on tree-roosting microbats living in forests, very little is known about tree-roosting microbats in southern Africa.

Yellow house bats and lesser yellow house bats are frequently found roosting in hollow spaces in the roofs of houses, rather than in the crevice-like situations used by many free-tailed bats. Occasionally, so-called 'cave bats', such as slit-faced, horseshoe and leaf-nosed bats, may be found in larger roofs or basements of houses.

Cave-dwelling bats (including natural and man-made cavities) form the largest aggregations of any mammal. As many as 20 million bats may occupy Bracken Caves in Texas. South Africa's largest known cave colony of bats at De Hoop Guano Cave, near Bredasdorp in the Western Cape, comprises some 300 000 bats. In southern Africa, cave-dwelling bats include the Egyptian fruit bat, horseshoe bats, leaf-nosed, and trident bats, slit-faced bats, long-fingered bats, and Temminck's hairy bat.

Bats use caves and other cavities to take advantage of the protection from predators and the stable microclimate (temperature and humidity). The microclimate preferences of bats change during the year. In summer, bats select warm and humid roosts to help their young to maintain their body temperature. In winter, hibernating bats select cooler roosts which allow them to decrease their body temperature (and consequently, their metabolic needs) so as to conserve energy while few insects are available. On the other hand, non-hibernating bats (e.g. common slit-faced bats) will select warmer roosts during winter to reduce energy expenses needed to maintain a constant body temperature.

The changing microclimate needs of different species of cave bats are well demonstrated by the colony of five common species of bats occupying De Hoop Guano Cave in the Western Cape (Figure 2). The different species, and even sexes of one species, use this cave at different times of the year and for different reasons. For example, during summer, a colony of up to 290 000 Schreibers's long-fingered bats are present. Males occupy cooler roosting sites near the cave entrance, while females and their young inhabit a domed 'maternity chamber' within the cave complex, where temperature and humidity are extremely high due to the combined body temperature of the bats, 31° C and 91% respectively. During winter, all the pregnant females of this species migrate 100–200 km from De Hoop Guano Cave to caves in the interior of the same province, where the lower ambient temperatures are necessary for hibernation. Males and non-pregnant females of this species remain at De Hoop during winter, without entering into deep hibernation. De Hoop is also an important maternity cave for Temminck's hairy bat, but the maternity roost of this species is somewhat cooler (21° C) and drier (85% relative humidity) than for Schreibers's long-fingered bat. Like Schreibers's long-fingered bats, the females and young migrate to colder caves during winter in order to hibernate.

On the other hand, the non-hibernating common slit-faced bat is present at De Hoop Guano Cave in autumn and winter but not in summer. Since it cannot hibernate or enter torpor, this species overwinters in a warm (30° C) roosting position, just beyond the 'maternity dome' of the Schreibers's long-fingered bat.

The two species of horseshoe bats, Cape horseshoe bat and Geoffroy's horseshoe bat, are present at De Hoop Guano Cave throughout the year, giving birth in December, but occur in much higher numbers during winter when Schreibers's long-fingered bats and Temminck's hairy bats are largely absent. They particularly favour a cold (19° C) roosting site just beyond the first bend in the cave, which results in most individuals being in a torpid state for much of winter. Although they are frequently torpid during the day in winter, horseshoe bats do not enter a deep prolonged hibernation at De Hoop Guano Cave, and actively forage throughout the year. This is because the winter rainfall climate of the Western Cape ensures a good supply of insects during that season. During summer the entrance site of the cave is also occupied by males and non-breeding females of Schreibers's long-fingered bats and Temminck's hairy bats.

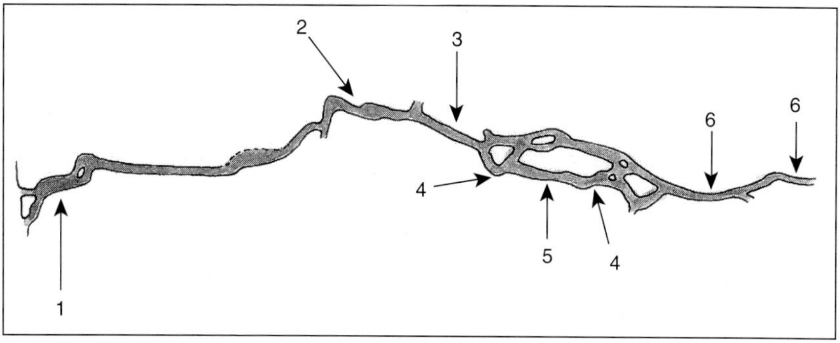

Figure 2 Cross section of De Hoop Guano Cave (after McDonald et al. 1990b)

1. Summer roosting site for non-breeding bats and winter roost for Schreibers's long-fingered bats, Geoffroy's and Cape horseshoe bats 19° C
2. Maternity chamber for Schreibers's long-fingered bats 31° C
3. Winter roost for common slit-faced bats 30° C
4. Maternity roosts for Temminck's hairy bats 21° C
5. Roost for horseshoe bats 21° C
6. Incidental roost sites for all five species 21° C

Crevice-roosting bats

Whilst also roosting in natural situations under the bark of trees, in fissures of hollow trees or in rocky crevices, crevice-roosting bats such as many pipistrelles, serotines, and many species of Molossidae (e.g. little, Egyptian, and Angolan free-tailed bats in southern Africa), typically exploit crevices in man-made structures such as buildings and bridges.

Suitable crevice-like spaces in buildings, usually the roof spaces, are readily exploited by bats. Although colonies of up to 2 500 little free-tailed bats have been estimated to occur in the roof of a sugar mill in KwaZulu-Natal, colonies of most house-dwelling bats are generally much smaller, from a few individuals to a few dozen. Details on how to deal with unwanted roof-dwelling bats are given in Chapter Nine, as well as further details on where and how to locate bat-roosting sites within the roof space, and the location of exit holes.

Specialised roost sites

Open structures, such as garages and outhouses, thatched game-viewing hides, and the eaves of buildings, are frequently used as night roosts. Many cave-dwelling bats carry their insect prey to these night roosts or feeding stations, which provide temporary shelter. Night-roosting bats can often be located by the piles of discarded insect parts below their feeding spots.

Apart from all the above-mentioned roosts, bats may be found roosting in a variety of other situations, such as in the nests of birds, e.g. sunbirds or weavers, which are favoured roost sites for woolly bats of the genus *Kerivoula*.

ECOLOGICAL AND ECONOMIC BENEFITS OF BATS

There is growing evidence that bats play vitally important ecological roles which may also have significant economic benefits. Bats are the major predators of night-flying insects. Co-operative Extension Farm Advisor from the University of California, Rachel Long, and colleagues found that agricultural pests featured prominently in the diet of bats occupying the fertile Sacramento Valley. In another study by Long and co-workers, reported in February 1999 in the *New Scientist* magazine, California pear farmers suffered crop losses of less than 5% due to the corn ear moth when a bat colony was situated within 2 km; when the bat colony was situated over 4 km away, crop losses of 60% were reported. In other words, the presence of sufficient numbers of bats reduced crop damage by 55%.

Every year, the adult moths of billions of corn earworms, fall armyworms, and other insects migrate in swarms from northern Mexico into Texas at altitudes of up to 3 km. These insect swarms cause massive crop losses across the southern and central United States, costing the US billions of dollars annually. Recent research by Gary McCracken and co-workers, using radar, weather balloons, and bat detectors has estimated that the 100 million Mexican free-tailed bats occupying Bracken Cave and other major caves in central Texas can eat approximately 1 000 tons of insects each night. If only 10% of the bat's diet were corn earworm moths (at 250 mg per moth), these bats would eat 340 million corn earworms each night, saving farmers millions of dollars.

Closer to home, South Africa's largest colony of 300 000 bats occupying the De Hoop Guano Cave assists farmers in the Bredasdorp area by consuming approximately 100 tons of insects annually, including many crop pests. With a growing emphasis on biological control and integrated pest management, more and more farmers are taking bats seriously, especially since a single colony of bats may consume millions of insects, including crop pests, each growing season. Instead of eradicating bat colonies from their farmhouses, farmers are now exploring ways of attracting bat colonies to their fields by erecting bat houses.

Considering that a single North American little brown bat can consume up to 1 200 mosquito-sized insects in an hour, or almost 5 000 in a single night, bats can be useful guests for controlling nuisance insects too. Mosquitoes have been found in the diet of certain species of bats worldwide, but much research is required to establish their importance in the diet of southern African bats.

Fruit bats also bring important ecological and economic benefits. Research sponsored by Bat Conservation International has shown that the seed dispersal and pollination activities of fruit-eating bats are vital to the survival of equatorial and tropical rain forests. Some 300 plant species in the Old World tropics (Africa and Asia) alone depend on bats for pollination or seed dispersal, or both, providing more than 450 economically important products valued in the hundreds of millions of US dollars annually. Certain bat-dependent trees, such as the baobab (whose white flowers may help attract bats at night), are ecologically crucial, supporting dozens of other species themselves. Extinction of baobabs resulting from the extinction of the epauletted fruit bats that pollinate them could trigger a cascade of linked extinctions.

Seeds dropped by tropical bats are estimated to contribute towards some 95% of forest regrowth on cleared land in the tropics.

WORKING WITH BATS

Capturing bats

Scientists commonly catch bats in the field by means of mistnets (Figure 3). Other collecting methods include harp traps (Figure 51), handnets, and taking them by hand from their day-time or night-time roosts. Mistnetting and collecting of bats can only be done with a permit from the authorised nature conservation agency, but locating of bat roosts need not involve any collection of specimens. Sometimes remains of bats' skeletons can be found in barn owl or eagle owl pellets (regurgitated balls of bones and fur found underneath the owl's nest); this can provide very useful data on bat distributions.

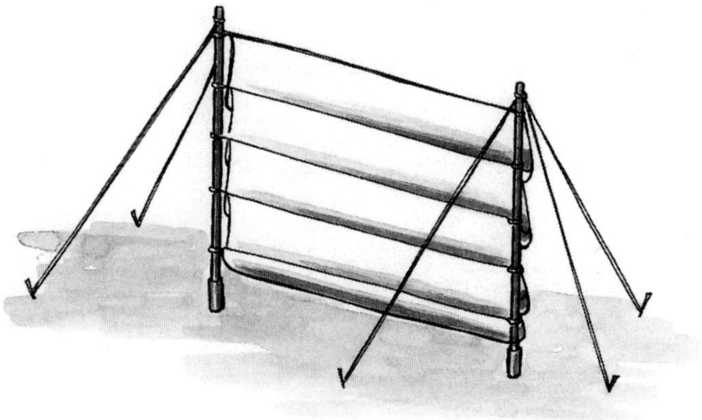

Figure 3 Mistnet structure (after Churchill 1998)

Handling, marking and measuring bats

Bats should only be handled by qualified scientists and bat-workers with the appropriate permits and rabies pre-exposure inoculation. Figure 4 shows the standard measurements which are typically taken on a bat. In the past, large numbers of bats have been banded, using metal alloy or plastic bands placed around the bat's forearm; however caution needs to be exercised, since bands can result in unacceptably high rates of injury and mortality in some species. Some workers advocate the use of alternative marking methods, such as necklaces or transponders. Transponders are minute transmitters which can be implanted under the bat's skin but which need to be detected subsequently at close range using specialised scanners, requiring the re-capture of released individuals.

With advancing technology and miniaturisation, very small radio transmitters are now available which can be fitted to all but the smallest bats (using a non-toxic surgical glue) to allow radio tracking. As a rule, the transmitter should weigh no more than 5% of the bat's body mass. Radio-tracking can be very useful for determining, for example, home ranges of individuals, foraging patterns, and the location of day roosts.

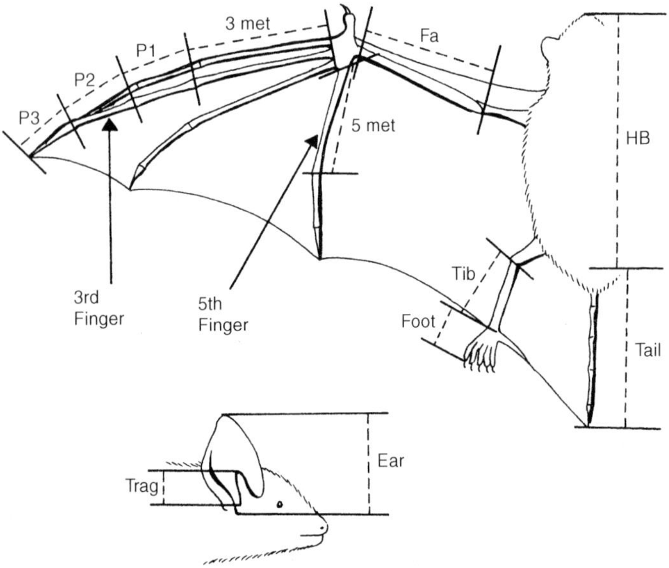

Figure 4 Measurements used in bat identification

Fa: forearm length
Tib: tibia or lower leg length
Ear: length of ear from notch to tip
Trag: length of tragus in the ear
3 met: length of metacarpal in third finger of wing
5 met: length of metacarpal in fifth finger of wing
Tail: tail length from tip to anus
HB: head and body length from nose tip to anus
Foot: hind foot length, heel to toe tips, excluding claws
P1: length of first phalanx of the third finger of the wing
P2: length of second phalanx of the third finger of the wing
P3: length of third phalanx of the third finger of the wing

Plate 1
Eidolon helvum
Straw-coloured flying fox

Plate 2
Epomophorus gambianus
Gambian epauletted fruit bat

Plate 3
Coleura afra
Sheath-tailed bat

Plate 4
Chaerephon pumilus
Little free-tailed bat

Plate 5
Mops condylurus
Angolan free-tailed bat

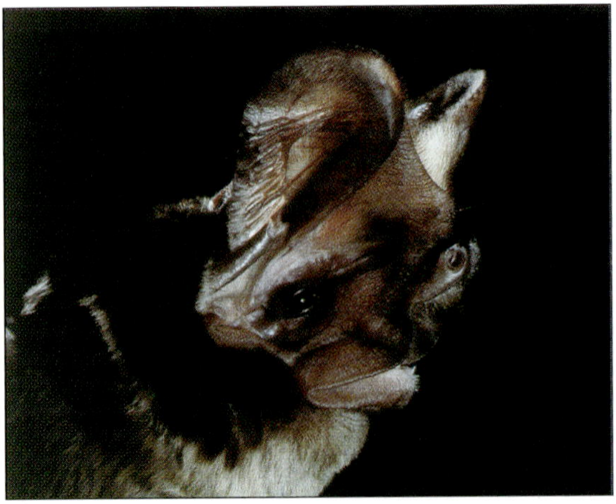

Plate 6
Otomops martiensseni
Large-eared free-tailed bat

Plate 7
Miniopterus inflatus
Greater long-fingered bat

Plate 8
Laephotis botswanae
Botswana long-eared bat

Plate 9
Myotis tricolor
Temminck's hairy bat

Plate 10
Pipistrellus kuhlii
Kuhl's pipistrelle

Plate 11
Pipistrellus rueppellii
Rüppell's bat

Plate 12
Scotophilus dinganii
Yellow house bat

Plate 13
Nycteris hispida
Hairy slit-faced bat

Plate 14
Nycteris thebaica
Common slit-faced bat

Plate 15
Rhinolophus landeri
Lander's horseshoe bat

Plate 16
Cloeotis percivali
Short-eared trident bat

Plate 17
Hipposideros commersoni
Commerson's leaf-nosed bat

Plate 18
Triaenops persicus
Persian leaf-nosed bat

Section Two

Identification guide to southern African bats

Chapter Four

How to use this guide

HISTORICAL PERSPECTIVE

The availability of identification keys and distribution maps for animals and plants is often taken for granted, overlooking the enormous collecting efforts of successive generations of pioneering natural historians. During the eighteenth and nineteenth centuries, when much of southern Africa fell under colonial rule of either Britain, Portugal, or Germany, early explorers and hunters such as Temminck and Wahlberg (to take just two examples with bats named after them) collected specimens which were then sent back to the major museums of Europe, such as the Natural History Museum in London (previously known as the British Museum of Natural History). Within the past few decades, mammalogists such as D.L. Harrison, R.W. Hayman, J.E. Hill, J. Kingdon, K.F. Koopman, D.R. Rosevear, and O. Thomas, have done an excellent job of making sense out of these invaluable historical collections and publishing the results in books and journals.

In the early twentieth century, local mammal collectors such as G.C. Shortridge, W.L. Sclater, and A. Roberts worked prodigiously to build up collections of southern African bats and other mammals, and this tradition has been continued in the latter part of the twentieth century by people such as R.H.N. Smithers and many others. An exhaustive survey of African mammals by the Smithsonian Institution during the 1960s and 1970s led to a reference work on *The mammals of Africa: An identification manual,* authored between 1971 and 1977 by J. Meester and H.W. Setzer, with the chapter on bats authored by R.W. Hayman and J.E. Hill. The reference provides the most up to date key for many areas of Africa, having being superceded to some extent in southern Africa by *Classification of southern African mammals* by J.A.J. Meester and colleagues (1986).

Fortunately, the efforts of these collectors have resulted in several major museum collections of southern African bats, the value of which increases with time, particularly more so in the light of their importance to national biodiversity conservation efforts, as explicitly recognised by the 1992 Convention on Biological Diversity (CBD). The distribution maps referred to in this section are based on museum specimens. These collections provide a vital scientific reference for the accurate identification of new specimens. Important southern African museum collections of bats include the Amathole Museum (King William's Town – previously the Kaffrarian Museum), Durban Natural Science Museum, MacGregor Museum (Kimberley), Natal Museum (Pietermaritzburg),

National Museum (Bloemfontein), South African Museum (Cape Town), Transvaal Museum (Pretoria), and the Zimbabwe Natural History Museum (Bulawayo).

In spite of some two centuries of past collecting, our knowledge of the basic aspects of bat biology, such as their distribution, remains very poor. There are just too few collectors and scientists relative to the high diversity of bats in our region. Here, the contribution of the enlightened and observant amateur bat-workers is vital. Significantly, four out of five southern African records of two rare species of free-tailed bats (*Tadarida lobata* and *T. ventralis*) were specimens found dead by the public and submitted to the local museum. A recent new species record for South Africa, Thomas's house bat, was the result of two amateur bat-workers responding to a pregnant female bat's distress calls (the bat later gave birth to twins and died in their care).

APPROACH TAKEN

Identifying bats must seem highly daunting to any beginner who may have tried wrestling with the identification keys provided in currently available mammal field guides and scientific texts. Detailed knowledge of anatomical terminology of the skull and teeth is required, and in some cases microscopic analysis of prepared museum specimens (skins and skulls) is necessary to accurately determine species.

Some simplification is necessary for the beginner. The illustrated identification key used in Chapter Five is based primarily on external characters which can be determined on the live bat in the field. The more difficult characters are illustrated to assist the layman. An expert should be consulted wherever possible (see Important contacts), as there is no substitute for expert identification using reference specimens stored in a museum.

In Chapter Six, species accounts are provided for all 74 species of bats found in southern Africa, accompanied by colour illustrations, distribution maps, sonograms, and tables of measurements. The order of families follows that given in the identification key in the following chapter. Within each family, genera and species are arranged alphabetically. Aspects covered by each species account are:

1) **Name**. The common and scientific names are given, together with the author and date of description of each genus and species.
2) **Measurements**. Mean and ranges are given for each species (males and females separately), for forearm length in millimetres (FA), and mass in grams based on published and unpublished sources cited under each table of measurements. In some cases, measurements were obtained by me from specimens in the following museum collections: Durban Natural Science Museum (DNSM), National Museum of Natural History, Paris (MNHN), or the Harrison Zoological Museum Institute, Sevenoaks, Britain (HZM). Unless otherwise indicated, data apply to specimens from southern Africa only. FA remains one of the most indispensable characters used in the identification of bats, and species ranges for FA are given both within tables of measurements and within the key in Chapter Five.

3) **Identification**. Main species characteristics, including features useful for identification, are explained in italics at the beginning of each account. Readers are referred back to the illustrated key in Chapter 5. Colour illustrations are included for 32 species to assist with identification.
4) **Habitat**. The habitat refers to the vegetation type (e.g. grassland, savanna, riverine forest), and certain habitat requirements such as close proximity of houses, caves, tunnels, rocky crevices, or water bodies. Information given is drawn primarily from regional texts such as Skinner and Smithers (1990) and references contained therein, or to more recent regional studies (Lynch 1994; Monadjem 1998; Taylor 1998), or data obtained from examination of museum specimen labels. Data from elsewhere in Africa are briefly noted in the absence of any data from southern Africa.
5) **Habits**. Any unique and distinguishing behavioural and social traits are mentioned here, based largely on the references listed above, or on personal field notes.
6) **Feeding**. Prey items, foraging behaviour and other relevant data are noted when available for southern Africa, or when data from elsewhere in Africa are likely to be true for southern Africa. General patterns and terms are described previously in Chapter Three. Information on prey items and feeding behaviour is based on a number of studies of southern African species which have generally described prey contents from the droppings of different species of bats, and related these findings to comparative observations of echolocation, wing morphology and feeding behaviour (Aldridge and Rautenbach 1987; Bowie et al. 1999; Jacobs 1999; and numerous papers by M.B. Fenton).
7) **Reproduction**. Gestation, litter size, birth season, and any other relevant details are noted when data are available for southern Africa, or when available only for other parts of Africa and are likely to be true for the subregion. General patterns are described in Chapter Three. Information given is based on numerous studies on reproduction of southern African species by R.T.F. Bernard; F.P.D. Cotterill; D.C.D and M. Happold; M. van der Merwe; L. Vivier; and others.
8) **Day roost**. Categories introduced in Chapter Three are used (i.e. foliage, hollow, crevice, or specialist roosts), together with more specific details where available. In the case of many cave-roosting species, night roosts (sheltered spots where bats bring items of food to eat) are also detailed.
9) **Call**. Wherever available, sonograms (time-frequency graphs) and/or brief descriptions of the echolocation calls are given. Descriptions include call structure, i.e. frequency modulated (FM) or constant frequency-dominated (CF); as well as the frequency range, maximum call duration, and the dominant frequency (i.e. the frequency at which signal strength is greatest). Data and sonograms for bat echolocation calls are invariably based on the recent study by Taylor (1999) and references cited therein.
10) **Distribution and status**. Generalised range maps are provided for the distribution of each species throughout southern Africa. Additional comments on

the distribution, commonness, or rarity of each species are given where appropriate. Where official protection status exists under provincial conservation ordinances, this is mentioned. The pest status of certain species is explained. For species listed in the *1996 IUCN Red List of Threatened Species*, the category of threat is given. The *1996 IUCN Red List of Threatened Species*, as well as a full explanation of the threat categories, can be consulted via the Internet at the following web page: **http://www.wcmc.org.uk/species/animals/**. A list of threatened southern African species, with explanation of IUCN threat categories, is given in Chapter Seven.

CLASSIFICATION AND MORPHOLOGICAL TERMS

Taxonomists use a hierarchy of names to classify all living organisms, from kingdom down to species and subspecies, usually termed the Linnaean hierarchy. Understanding this hierarchy of names is very useful as it allows one to group information about large numbers of species. For example, all members of the family of free-tailed bats (Molossidae) have the tails projecting free from the tail membrane. To illustrate the different levels of classification, a simple example is given below.

Kingdom	Animal
Phylum	Chordata (backboned animals and their immediate relatives)
Class	Mammalia
Order	Chiroptera
Suborder	Megachiroptera
Family	Pteropodidae
Genus	*Epomophorus*
Species	*wahlbergi*

Generally species are referred to by their genus (with the first letter upper case) and species (all lower case), names only, in italics, e.g. *Epomophorus wahlbergi* (Wahlberg's epauletted fruit bat).

The detailed classification and identification of bats at the species level is sometimes complex and relies largely on dental and cranial characteristics. However, it is a relatively simple task to identify the family to which a species belongs simply by observing the relationship between the tail and the tail membrane, and also the facial features. These features are emphasised in the identification keys provided in Chapter Five.

Using identification keys takes some practice but is really fairly simple. Keys consist of a sequence of 'couplets' (two-choice questions) which are followed systematically from top to bottom. Depending on your answer to each question, you are either directed to another couplet, or given your final answer as to the identification of the specimen you are holding.

Answering the questions to the keys requires some familiarity with anatomical terms for bats. For example, in order to identify a long-fingered bat you need to know which is the second phalanx of the third digit. The diagram following (Figure 5) gives some of the anatomical terms relating to bats, and will help to orientate you as to where to start counting fingers (i.e. from the claw, which is all that remains of digit one, the thumb). The Glossary provides further assistance with unfamiliar terms, as do the illustrations which accompany the key.

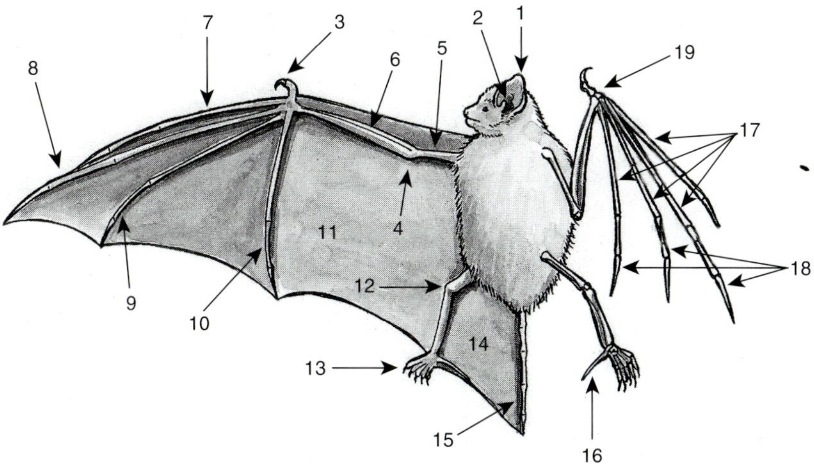

Figure 5 Parts of a bat (*Pipistrellus*)

1. ear
2. tragus
3. thumb/1st finger
4. elbow
5. upper arm
6. forearm
7. 2nd finger
8. 3rd finger
9. 4th finger
10. 5th finger
11. wing membrane
12. knee
13. foot
14. tail membrane
15. tail
16. calcar (some species)
17. metacarpals
18. phalanges
19. wrist

Chapter Five

Key to the families, genera and species

The following keys were derived mostly from the 1986 reference, *Classification of southern African mammals*, by J.A.J. Meester and co-workers. This reference in turn used and modified keys from earlier sources, most frequently, Hayman and Hill's (1977) chapter in J. Meester and H.W. Setzer's *Mammals of Africa: An identification guide*. Most of the keys to the Family Molossidae are based on a 1975 publication by K.F. Koopman on *Bats of the Sudan*. I have modified the keys presented in this book by emphasising external characters which may be useful in the field, rather than just relying on skull and dental characters. Information on echolocation calls is also included in the keys for the first time. Unpublished field observations and suggestions of F.P.D. Cotterill and M.B. Fenton (pers. comm.) were invaluable in improving certain of the species keys, particularly for the genera, *Nycteris, Tadarida* and *Rhinolophus*.

1. Claw on second finger, FA 66–115 mm long; face dog-like (**Suborder MEGACHIROPTERA, Family PTEROPODIDAE: fruit bats**) 2

Figure 6 Head of Wahlberg's epauletted fruit bat

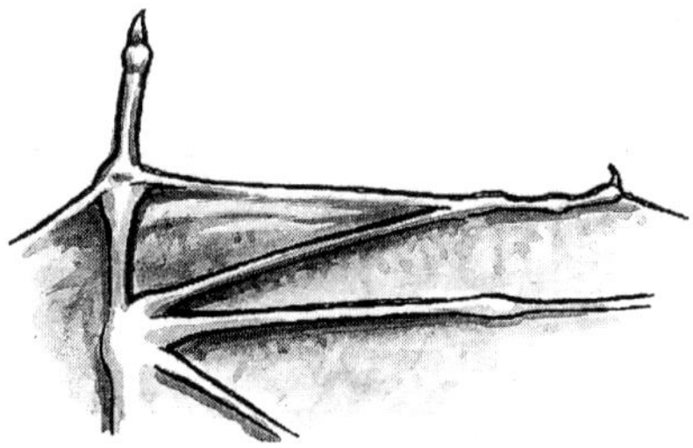

Figure 7 Wing of Wahlberg's epauletted fruit bat showing two claws

1' Second digit without claw; FA <66 mm long; face not dog-like (**Suborder MICROCHIROPTERA, six Families**) 9

2. FA >110 mm long (110–130 mm); colour generally tawny, with dorsal fur restricted to a narrow median band, sharply demarcated from naked wing membranes .... **Straw-coloured flying fox**, *Eidolon helvum*

2' FA <110mm long 3

3. Ears with white basal tufts; tail absent 4

3' Ears without white basal tufts; reduced tail present 7

Figure 8 Tail area (from skin) of (a) Wahlberg's epauletted fruit bat and (b) Egyptian fruit bat

4. Rousettes of males particularly developed, yellowish in colour; face shortened, shorter than braincase; distribution in southern Africa limited to Chobe River in Botswana; FA 80–92 mm
.... **Dobson's fruit bat**, *Epomops dobsonii*

4' Rousettes less conspicuous, white-coloured; face longer than braincase; widespread distribution; FA 68–95 mm
(**Epauletted fruit bats**, *Epomophorus*) 5

5. One palatal ridge behind last molar; FA 68–95 mm
.... **Wahlberg's epauletted fruit bat**, *Epomophorus wahlbergi*

5' Two palatal ridges present behind last molar; FA 75–94 mm 6

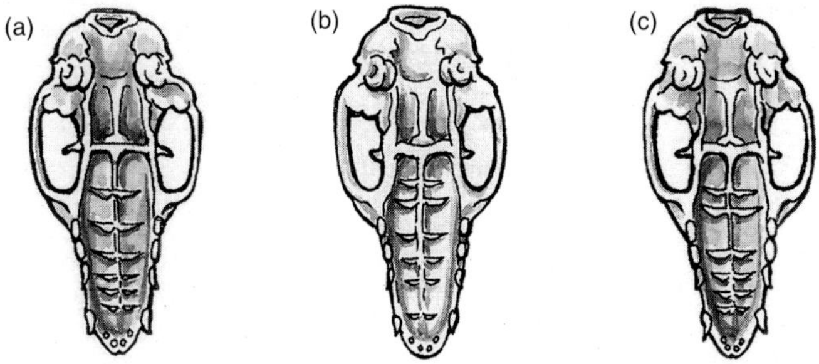

Figure 9 Palatal ridges compared in (a) Wahlberg's epauletted fruit bat (b) Gambian epauletted fruit bat (c) Angolan epauletted fruit bat

6. 4th palatal ridge halfway between 3rd and 5th; widespread distribution; FA 75–88 mm .... **Gambian epauletted fruit bat**, *Epomophorus gambianus*

6' 4th palatal ridge closer to 3rd than 5th; southern African distribution restricted to southwestern Angola and Ovamboland in northern Namibia; FA 81–94 mm .... **Angolan epauletted fruit bat**, *Epomophorus angolensis*

7. Wing membranes arise from first toe; hair short and sleek; FA >89 mm (89–96 mm); ruff of stiff hairs absent from throat of males; tail fairly robust .... **Egyptian fruit bat**, *Rousettus egyptiacus*

7' Wing membranes arise from second toe; hair longer and coarser; FA <90 mm; ruff of stiff hairs present on throat of males, tail relatively shorter, less robust 8

8. Size smaller, FA <80 mm (66–76 mm); distribution in southern Africa limited to Haroni-Rusitu confluence in eastern Zimbabwe .... **Bergmans's small fruit bat**, *Myonycteris relicta*

8' Size larger, FA >80 mm (82–90 mm); occurring more widely in the Eastern Highlands of Zimbabwe and central Mozambique .... **Bocage's fruit bat**, *Lissonycteris angolensis*

9. Tail ends in T-shaped cartilage; face with deep slit (**Family NYCTERIDAE, genus *Nycteris*: slit-faced bats**) 10

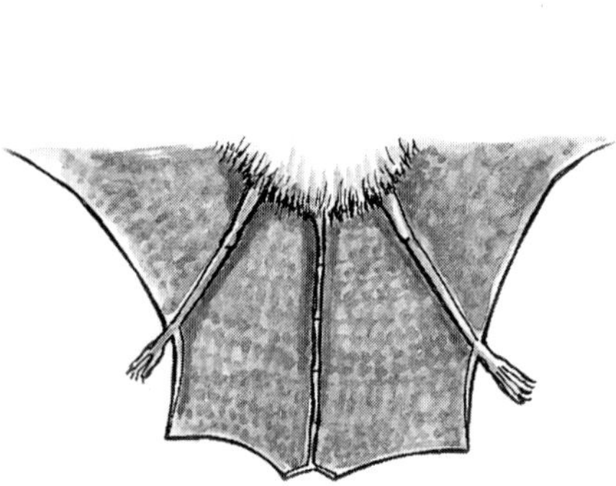

Figure 10 Tail area (from skin) of common slit-faced bat

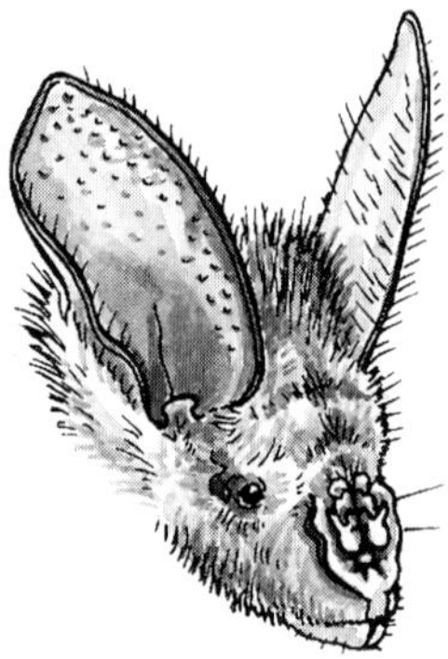

Figure 11 Face of hairy slit-faced bat showing facial slit

9' Tail without T-shaped cartilage; face without deep slit 14

10. Larger in size, FA 57–66 mm, ear 28–35; mass over 30 g .... **Large slit-faced bat**, *Nycteris grandis*

10' Smaller in size, FA 37–52 mm, ear 18–37 mm; mass less than 20 g........... 11

11. Tragus has a distinctive 'semi-lunate' (half-moon) shape; FA 36–50 mm, ear 18–34 mm 12

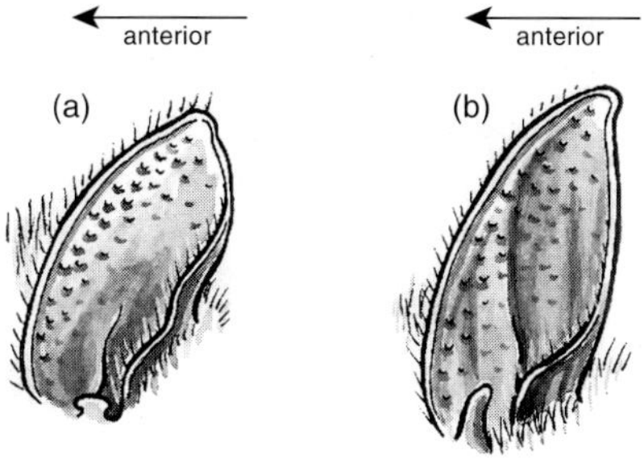

Figure 12 Tragus shapes of (a) hairy slit-faced bat (semi-lunate tragus) and (b) common slit-faced bat (pyriform tragus)

11' Tragus has a distinctive 'pyriform'(finger-like) shape; FA 42–52 mm, ear 28–37 mm .... **Common slit-faced bat**, *Nycteris thebaica*

12. Ears shorter, 18–25 mm; upper incisors trifid (three-lobed); FA 36–45 mm .... **Hairy slit-faced bat**, *Nycteris hispida*

12' Ears longer, 28–34 mm; upper incisors bifid (two-lobed); FA 37–50 mm 13

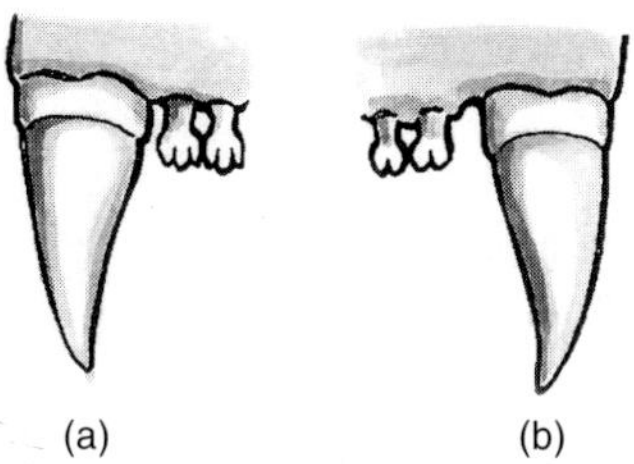

Figure 13 Trifid (a) versus bifid (b) incisors of slit-faced bats

13. General colour greyish; smaller, FA 37–42 mm, ear 29–34 mm .... **Wood's slit-faced bat**, *Nycteris woodi*

13' General colour brownish; larger, FA 45–50 mm, ear 28–34 mm .... **Greater slit-faced bat**, *Nycteris macrotis*

14. Tail extends beyond the edge of, or protrudes through, the interfemoral membrane 15

14' Tail almost entirely enclosed within interfemoral membrane (at most two or three vertebrae not included) 31

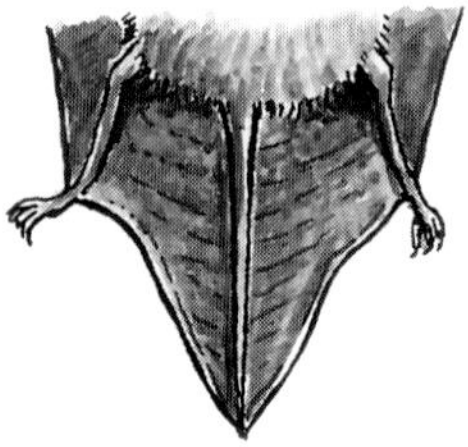

Figure 14 Tail area (from skin) of Schreibers's long-fingered bat

15. Free terminal portion of tail protrudes through middle of upper surface of tail membrane
(**Family EMBALLONURIDAE: sheath-tailed bats**) 16

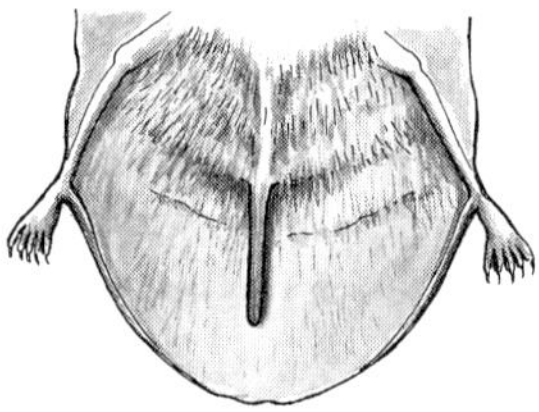

Figure 15 Tail area (from skin) of Mauritian tomb bat

15' Free terminal portion of tail extends considerably beyond hind margin of tail membrane
(**Family MOLOSSIDAE: free-tailed bats**) 18

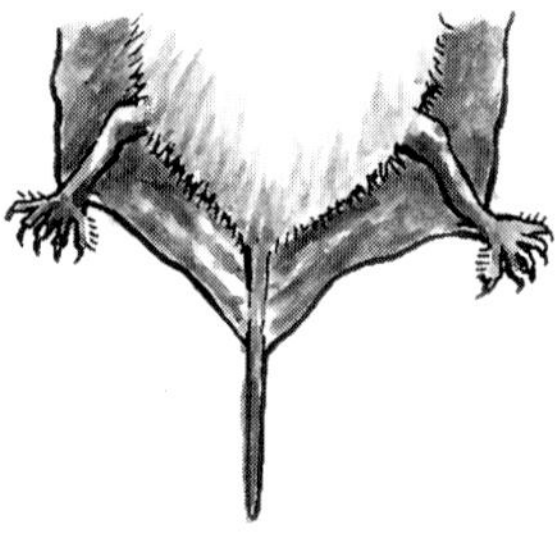

Figure 16 Tail area (from skin) of little free-tailed bat

16. Size smaller, FA <50 mm (46–49 mm); distribution in southern Africa restricted to Zambezi River in Mozambique
.... **Sheath-tailed bat**, *Coleura afra*

16' Size larger, FA >60 mm; widespread distribution in southern Africa (**Tomb bats**, *Taphozous*) 17

17. Colour grizzled-grey above and pure white below; gular throat pouch present in males; FA 60–66 mm, mean 63 mm .... **Mauritian tomb bat**, *Taphozous mauritianus*

17' Dark brown colour above, paler below but not white; no gular pouch in males; FA 62–63 mm, mean 62 mm .... **Egyptian tomb bat**, *Taphozous perforatus*

18. Size very large, FA 62–73 mm, ears very large, 38–40 mm, meeting on extended snout; conspicuous paler shoulder band separating darker anterior and posterior upper regions .... **Large-eared free-tailed bat**, *Otomops martiensseni*

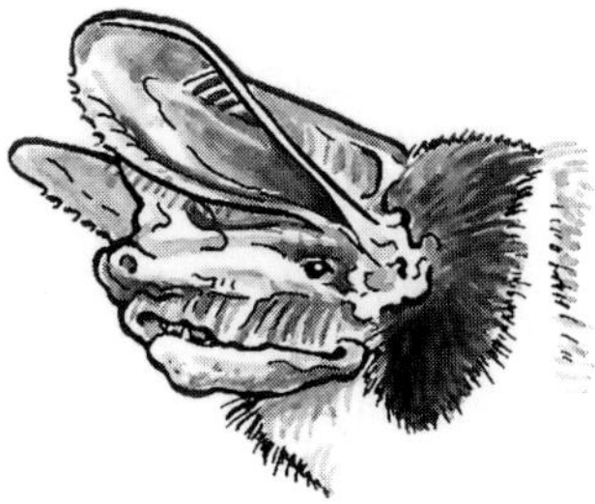

Figure 17 Head of large-eared free-tailed bat

18' Not combining above characters 19

19. Ears separate (**Genus *Mormopterus***) 20

19' Ears joined 21

20. Head conspicuously flattened, greatest braincase depth not more than a third of the greatest width; 'warty' muzzle; main 'brush' of hairs above nose; occurring widely in drier, rocky areas of the Western Cape, Northern Cape and Northern Province, Namibia, Botswana and southern Zimbabwe; FA 44 mm .... **Flat-headed free-tailed bat**, *Mormopterus petrophilus*

20' Skull not as conspicuously flattened; muzzle not warty; no brush of hairs above nose; known from southern Africa from just two specimens collected in Durban over 150 years ago .... **Natal free-tailed bat**, *Mormopterus acetabulosus*

21. Size smaller, FA 36–43 mm, or if larger than 43 mm, almost always having one or more of the following: 1) a dark blackish 'beard' on the throat, 2)

characteristic white flecks or short stripes on head and shoulders, or 3) conspicuous lateral band of white fur running lengthwise along underneath of the wing membrane where it joins the body; greatest skull length 16–21 mm; conspicuous crest of stiff hairs usually present on forehead of males; large ears usually joined above bridge of nose by a band of folded skin; anterior notch ('emargination') of palate usually reduced and separated from the palatal foramina by a bony bar, or, if including palatal foramina, width of emargination less than diameter of upper incisor
(**Wrinkle-lipped bats**, *Chaerephon*) 22

Figure 18 Head of little free-tailed bat

21' Size larger; FA 43–66 mm, greatest skull length 18–29 mm; no conspicuous tuft of fur on forehead of males; anterior palate with or without emargination; ears joined with band of skin or not 26

22. Size larger, FA 43–51 mm; greatest skull length 18–21 mm 23

22' Size smaller, FA 34–42 mm; greatest skull length 15–18 mm 25

23. White spots, stripes, or lateral band; anterior palate with or without conspicuous emargination; blackish throat patch absent; FA 44–51 mm; greatest skull length 18–21 mm 24

23' Without white markings or conspicuous lateral band; blackish throat patch present; colour umber above and below; FA 43–47.5 mm; greatest skull length 18–20 mm; anterior palate with emargination
.... **Ansorge's free-tailed bat**, *Chaerephon ansorgei*

24. Colour umber brown with variable pattern of white spots or short stripes arranged laterally on crown and sometimes also on flanks and shoulders; FA 46–51 mm; greatest skull length 19–21 mm; anterior palate with conspicuous emargination
.... **Spotted free-tailed bat**, *Chaerephon bivittatus*

24' No spots or stripes, but conspicuous lateral band of long, white hairs below, along flanks at junction with wing membrane, contrasting with very dark brown colour of body above and below; FA 44–50 mm; greatest skull length 18–19 mm; anterior palate without conspicuous emargination
.... **Nigerian free-tailed bat**, *Chaerephon nigeriae*

25. Males with long (13–14 mm) bicoloured postaural crest (dark basally, pale apically); light wings; colour below light with whitish mid-ventral band down to mid-belly; FA 34–39 mm; greatest skull length 16–17 mm .... **Pale free-tailed bat**, *Chaerephon chapini*

25' Males with short (10 mm) unicoloured postaural crest; wings light or dark; colour below variable, with or without white belly-stripe on the median area; FA 37–42 mm; greatest skull length 16–18 mm .... **Little free-tailed bat**, *Chaerephon pumilus*

26. Paler band of short hairs in front of shoulders, giving appearance of being naked; ears are joined above bridge of nose by band of skin; colour variable above but always conspicuously paler below, with variable degree of white colouring ventrally; third upper molar reduced to V-shape, or, in *Mops condylurus*, with third 'leg' noticeably shorter in length than the second; anterior emargination of palate not well developed, with palatal foramina separated by bony bar; FA 45–66 mm, greatest skull length 20–29 mm (**Mops free-tailed bats**, *Mops*) 27

26' No naked band in front of shoulders; ears just meet across bridge of nose, but not conspicuously joined by a band of skin; colour variable above, but underparts tend to be same colour or only slightly paler, though white markings may be present; third upper molar N-shaped, with third 'leg' almost as long as second; anterior emargination of palate well developed, width of emargination greater than diameter of upper incisor; FA 45–66 mm, greatest skull length 18–26 mm (**Guano bats**, *Tadarida*) 28

27. Smaller in size, FA 43–49 mm, greatest skull length 20–23 mm .... **Angolan free-tailed bat**, *Mops condylurus*

27' Larger in size, FA 58–66 mm, greatest skull length 27–29 mm .... **Midas free-tailed bat**, *Mops midas*

28. Smaller-sized, FA 45–54 mm, greatest skull length 18–19 mm; lips wrinkled .... **Egyptian free-tailed bat**, *Tadarida aegyptiaca*

28' Larger-sized, FA 57–66 mm, greatest skull length 22–26 mm; lips not wrinkled 29

29. Ears extremely large (c. 50 mm), protruding beyond snout when laid forward; white patch of fur between shoulders; interdental palatal length more than twice palatal width at anterior end; FA 59–62.5 mm; greatest skull length 22.5–23.5 mm .... **Big-eared free-tailed bat**, *Tadarida lobata*

29' Ears not as large or extended to same degree (19–25 mm); interdental palatal less than twice palatal width at anterior end; FA 57–66 mm; greatest skull length 22.0–25.7 mm 30

30. Larger, FA 61–66 mm, greatest skull length 23.5–25.7 mm .... **Giant African free-tailed bat**, *Tadarida ventralis*

30' Smaller, FA 57–59 mm, greatest skull length 22.0–23.0 mm .... **Madagascar large free-tailed bat**, *Tadarida fulminans*

31. Muzzle with nose-leaves 32

31' Muzzle without nose-leaves (**Family VESPERTILIONIDAE: plain-faced bats**) 45

32. Posterior leaflet of noseleaf with upper edge either elliptical in outline or tridentate (three-pronged) (**Family HIPPOSIDERIDAE: leaf-nosed and trident bats**) 33

Figure 19 Face of Sundevall's leaf-nosed bat

Figure 20 Face of Persian leaf-nosed bat

32' Posterior leaflet of noseleaf triangular with erect, pointed tip (lancet) (**Family RHINOLOPHIDAE: horseshoe bats**) 36

Figure 21 Face of bushveld horseshoe bat

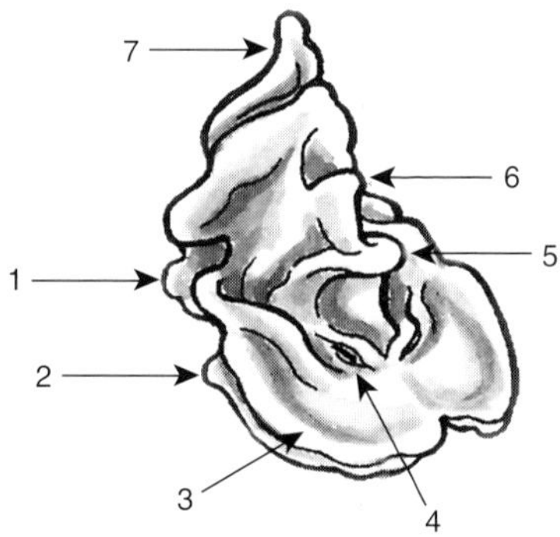

Figure 22 Terminology of the structures of the horseshoe in horseshoe bats

1. posterior noseleaf
2. secondary leaflet
3. anterior noseleaf
4. nostril
5. sella
6. connecting process
7. lancet

33. Posterior leaflet of noseleaf with upper edge elliptical in outline; size larger, FA >40 mm
(**Leaf-nosed bats**, *Hipposideros*) 34

33' Posterior leaflet of noseleaf with tridentate (three-pronged) shape; size smaller, FA 31–35 mm 35

34. Much larger-sized, FA 100–108 mm; fur short, colour below pure white .... **Commerson's leaf-nosed bat**, *Hipposideros commersoni*

34' Much smaller-sized, FA 45–50 mm; fur long and fluffy, brownish-grey in colour .... **Sundevall's leaf-nosed bat**, *Hipposideros caffer*

35. Size very small, FA 31–35 mm; ears rounded
.... **Short-eared trident bat**, *Cloeotis percivali*

35' Size larger, FA 48–54 mm; ears pointed
.... **Persian leaf-nosed bat**, *Triaenops persicus*

36. Face and/or lateral margins of sella covered with long hairs; greatest width of horseshoe over 9 mm; FA 48–67 mm; CF component of echolocation call 55 kHz or less 37

36' Sella naked; greatest width of horseshoe less than 9 mm; FA 40–57 mm; CF component of echolocation call 80 kHz or more 38

37. Size larger, FA 62–67 mm; echolocation call has CF frequency of 37–46 kHz
.... **Hildebrandt's horseshoe bat**, *Rhinolophus hildebrandtii*

37' Size smaller, FA 48–51 mm; echolocation call has CF frequency of 55 kHz
.... **Rüppell's horseshoe bat**, *Rhinolophus fumigatus*

38. Connecting process (Figure 22) rises to erect point or high and narrow horn; FA 41–47 mm 39

38' Connecting process blunt and rounded (Figure 22); FA 40–57 mm 40

39. The concave lancet (Figure 22) from the posterior noseleaf has a narrowly pointed shape, sella not rising to a very sharp narrow tip; fur shorter and darker; first phalanx of fourth finger (Figure 23) noticeably shortened in relation to metacarpal length; axillary (under-arm) tuft of stiff hairs in males; FA 41–43 mm .... **Lander's horseshoe bat**, *Rhinolophus landeri*

39' Margins of lancet not concave, forming equilateral triangle, sella rising to a very sharp narrow tip; fur longer and paler; first phalanx of fourth finger not noticeably shortened in relation to metacarpal length; no axillary tuft of stiff hairs in males; FA 41–47 mm
.... **Peak-saddle horseshoe bat**, *Rhinolophus blasii*

Figure 23 Wing showing shortened first phalanx of fourth finger of Lander's horseshoe bat

40. Canine and posterior premolar in contact, anterior premolar exterior to toothrow and minute, invisible to naked eye or hand lens; echolocation call has CF component at 86 or 94 kHz; FA 43–57 mm 41

40' Canine and posterior premolar not in contact, separated by anterior premolar which is just visible to naked eye, or with hand lens; echolocation call has CF component of 80, 83 or >100 kHz; FA 40–52 mm 42

41. Larger, FA 50–57 mm; echolocation call has CF component at 94 kHz .... **Geoffroy's horseshoe bat**, *Rhinolophus clivosus*

41' Smaller, FA 43–49 mm; echolocation call has CF component at 86 kHz .... **Darling's horseshoe bat**, *Rhinolophus darlingi*

42. Larger, FA 48–52 mm; distribution limited to the Western Cape and Eastern Cape; echolocation call has CF component at 90 kHz .... **Cape horseshoe bat**, *Rhinolophus capensis*

42' Smaller, FA 40–46 mm; echolocation call has CF component of 83 kHz, or >100 kHz 43

43. Ears longer, >22 mm; CF component of echolocation call 83 kHz; larger, FA 42–46 mm .... **Bushveld horseshoe bat**, *Rhinolophus simulator*

43' Ears shorter, <22 mm; CF component of echolocation call >83 kHz; smaller, FA 40–44 mm 44

44. Front edge of connecting process almost concave; lancet with sides almost convex; dorsal fur long and whitish at base, tips grey-brown, ventrally pure white; occurs only in more arid areas of Botswana, Namibia and the Northern Cape; call unrecorded but almost certainly >100 kHz, based on small body size .... **Dent's horseshoe bat**, *Rhinolophus denti*

44' Front edge of connecting process convex; lancet with sides almost concave, tip pointed; fur nearly unicoloured, not whitish at base; occurs only in savanna areas of Zimbabwe and eastern South Africa; CF frequency 110–115 kHz .... **Swinny's horseshoe bat**, *Rhinolophus swinnyi*

45. Second phalanx of third digit about three times as long as first (Figure 24) (**Subfamily MINIOPTERINAE: long-fingered bats**) 46

Figure 24 Wing showing elongated second phalanx of third finger of Schreibers's long-fingered bat

45' Second phalanx of third digit less than three times as long as first 48

46. Larger-sized, FA 45–50 mm; mass averages 15 g; greatest skull length exceeds 16 mm .... **Greater long-fingered bat**, *Miniopterus inflatus*

46' Smaller-sized, FA 42–47 mm; mass 9–16 g; greatest skull length 14–15 mm 47

47. Smaller, FA 42–44 mm, mass 9–11 g; skull c. 14 mm .... **Lesser long-fingered bat**, *Miniopterus fraterculus*

47' Larger, FA 42–47 mm, mass 9–16 g; skull c. 15 mm .... **Schreibers's long-fingered bat**, *Miniopterus schreibersii*

48. Ears funnel-shaped, with deep emargination below tip; tragus long, narrow, sharply pointed (Figure 25); braincase high and rounded; long, soft, erect, woolly hair, with curled tips, giving grizzled appearance (**Subfamily KERIVOULINAE: woolly bats**) 49

48' Ears not funnel-shaped, without deep emargination below tip; tragus not sharply pointed; braincase not high and rounded (**Subfamily VESPERTILIONINAE**) 50

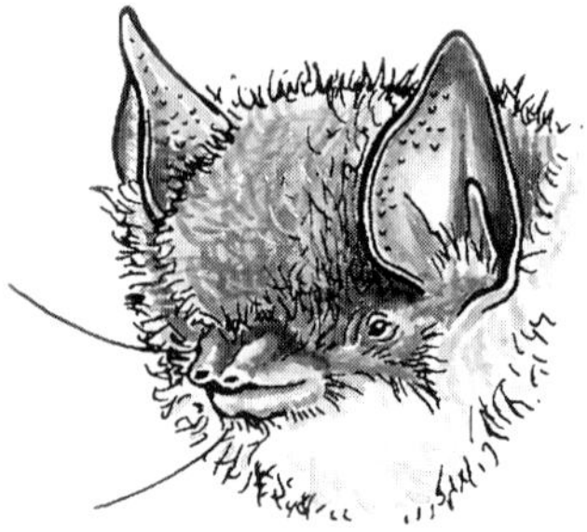

Figure 25 Face of Damara woolly bat

49. Size larger, FA 34–39 mm; greatest skull length 15 mm or more; colour of upper parts reddish-chestnut, underparts whitish or with buffy tinge .... **Damara woolly bat**, *Kerivoula argentata*

49' Size smaller; FA 30–36 mm; greatest skull length 13.5 mm or less; colour of upper parts brownish, may be lightly grizzled with greyish-white or white, underparts brown to greyish-brown to greyish-white to white .... **Lesser woolly bat**, *Kerivoula lanosa*

50. Wing extremely narrow, third digit shorter than head and body length; skull flattened; tragus extremely short and blunt; hair very short .... **Moloney's flat-headed bat**, *Mimetillus moloneyi*

50' Wing not extremely narrow, third digit longer than head and body; skull not flattened; tragus not extremely short and blunt; hair not so short 51

51. Fur coppery-reddish or yellowish to greyish-yellow in colour above and long and fluffy in texture; six upper cheekteeth and six lower cheekteeth on each side; elongated muzzle (**Hairy bats**, *Myotis*) 52

51' Fur not coppery-reddish in colour or long fluffy in texture; fewer than 6 upper and lower cheekteeth; muzzle not elongated 56

52. Size larger, FA >52 mm; distinctive pattern of contrasting black and coppery markings on wings and tail membrane .... **Welwitsch's hairy bat**, *Myotis welwitschii*

52' Size smaller, FA usually <52 mm (but may be up to 54 mm); colour uniform coppery-red or yellowish, without black markings on wings or tail membrane 53

53. Wings without glands; size larger, FA 41–54 mm 54

53' Wings with glands; size smaller, FA 32–35 mm 55

54. Size larger, FA 47–52 mm .... **Temminck's hairy bat**, *Myotis tricolor*

54' Size smaller, FA <47 mm .... **Rufous hairy bat**, *Myotis bocagei*

55. Size smaller, FA 32–33 mm; upper parts dull greyish-yellow
.... **Angola wing-gland bat**, *Myotis seabrai*

55' Size larger, FA >34 mm; upper parts honey yellow
.... **Lesueur's wing-gland bat**, *Myotis lesueuri*

56. Wing membranes with conspicuous reticulated pattern; FA 45 mm
.... **Butterfly bat**, *Chalinolobus variegatus*

56' Wing membranes without conspicuous reticulated pattern 57

57. Ear length 18 mm or more, about half length of forearm
(**Long-eared bats**, *Laephotis*) 58

57' Ear length <18 mm, or much less than half of forearm 60

58. Larger, FA >37 mm, greatest skull length >16 mm, ear length >19 mm 59

58' Smaller, FA <37 mm (33–37 mm), greatest skull length <15 mm, ear length <19 mm .... **Botswana long-eared bat**, *Laephotis botswanae*

59. Ear length 24–25 mm; FA 38–39 mm; colour light buffy brown; distribution restricted to the Namib Desert
.... **Namib long-eared bat**, *Laephotis namibensis*

59' Ear length 19–22 mm; FA 36–41 mm; colour buffy brown; occurs in the Western Cape, Free State and KwaZulu-Natal
.... **De Winton's long-eared bat**, *Laephotis wintoni*

60. Size larger, FA >43 mm 61

60' Size smaller, FA <43 mm 64

61. Tragus characteristically long and fine-pointed (Figure 26), the outer margin convex with a lobe towards its base; pronounced muzzle; dorsal fur soft with sheen, yellowish-brown coloured; underparts yellow, greyish-white or white; one pair of upper incisors present; FA 43–80 mm
(**House bats**, *Scotophilus*) 62

61' Tragus shorter, knife-shaped (Figure 27); muzzle not pronounced; dorsal fur rougher, without sheen, fawn to dark-brown coloured; ventral colour greyish; two pairs of upper incisors present, although outer incisors tiny, only visible with magnifying lens; FA 46–50 mm
.... **Hottentot serotine bat**, *Eptesicus hottentotus*

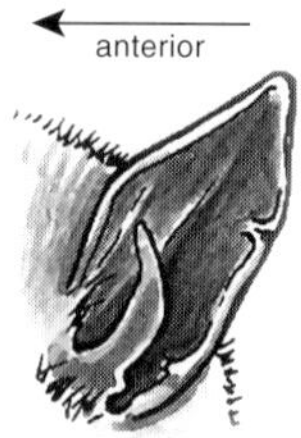

Figure 26 Ear and tragus of yellow house bat

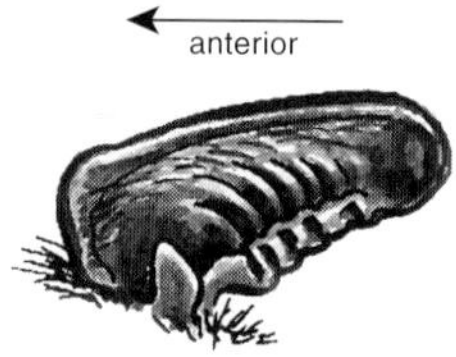

Figure 27 Ear and tragus of Hottentot serotine bat

62. Larger-sized, FA 70–80 mm, greatest skull length >30 mm .... **Giant yellow house bat**, *Scotophilus nigrita*

62' Smaller-sized, FA 43–65 mm, greatest skull length 16–23 mm 63

63. Size smaller, FA 44–52 mm, greatest skull length 16–20 mm; adults with whitish belly .... **Lesser yellow house bat**, *Scotophilus viridis*

63' Size larger, FA 50–65 mm; greatest skull length 20–23 mm; adults with yellow belly .... **Yellow house bat**, *Scotophilus dinganii*

64. Distinctive pale fawn dorsal colour, slightly paler below, contrasting with much darker brown coloured wings and interfemoral membrane; characteristic tragus shape, indented on the inner surface and with two small lobes on the outer edge (Figure 28); FA 28–33 mm .... **Schlieffen's bat**, *Nycticeinops schlieffeni*

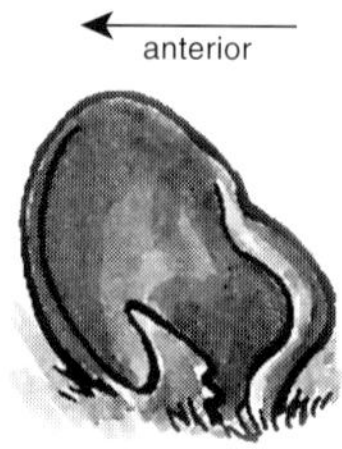

Figure 28 Ear and tragus of Schlieffen's bat

64' Colour and tragus shape not as above 65

65. Single, pair of prominent, procumbent upper incisors; rostrum broad, c. 7 mm, about 50% of greatest skull length; penis elongated, reaching 25% of head and body length; FA 30–31 mm (**Lesser house bats**, *Scotoecus*) 66

65' Two pairs upper incisors, outer much smaller than inner; rostrum not broad, c. 5–6 mm, or 40% of greatest skull length; penis not so elongated, less than 25% of head and body (**Pipistrelles**, *Pipistrellus*) 67

66. Pale-winged, four or five upper cheekteeth .... **Thomas's lesser house bat**, *Scotoecus albofuscus*

66' Dark-winged, four upper cheekteeth .... **Dark-winged lesser house bat**, *Scotoecus hirundo*

67. Underparts pure white; upper, inner incisors deeply bifid (i.e. divided) .... **Rüppell's bat**, *Pipistrellus rueppellii*

67' Underparts brown or grey; upper, inner incisors not deeply bifid 68

68. Five upper cheekteeth, including a tiny anterior premolar, creating a gap between canine and posterior premolar which may be visible with a magnifying lens in the field 69

68' Four upper cheekteeth, lacking tiny anterior premolar; canine and posterior premolar closely pressed together (Subgenus *Neoromicia*, previously grouped in the genus *Eptesicus*) 72

69. Tragus characteristically 'hatchet-shaped' with sharp obtuse angle in posterior edge; outer upper incisor more than half length of inner upper incisor; forehead strongly concave; typically found roosting in unfurling banana or *Strelizia* leaves; FA 29–38 mm .... **Banana bat**, *Pipistrellus nanus*

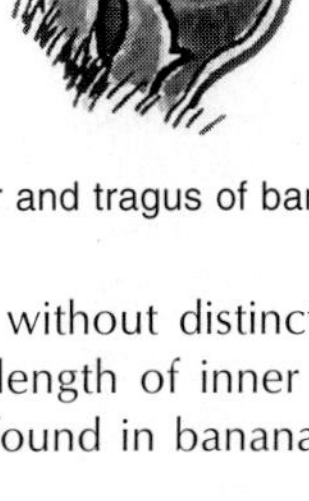

Figure 29 Ear and tragus of banana bat

69' Tragus not 'hatchet-shaped', without distinct angle in posterior edge; outer upper incisor less than half length of inner upper incisor; forehead usually flat or weakly concave, not found in banana or *Strelizia* leaves 70

70. Interdental palate as broad as long; forehead flat; total length of skull 12 mm or less; FA 27–29 mm .... **Rusty bat**, *Pipistrellus rusticus*

Figure 30 Ear and tragus of rusty bat

70' Interdental palate longer than broad; forehead at least weakly concave; total length of skull more than 12 mm; FA 29–35 mm 71

71. Posterior upper incisor barely extending above cingulum of anterior incisor; anterior upper premolar a pointed tooth rising above cingulum of canine; maxillary toothrow more than 4.5 mm; FA 31–35 mm .... **Kuhl's pipistrelle**, *Pipistrellus kuhlii*

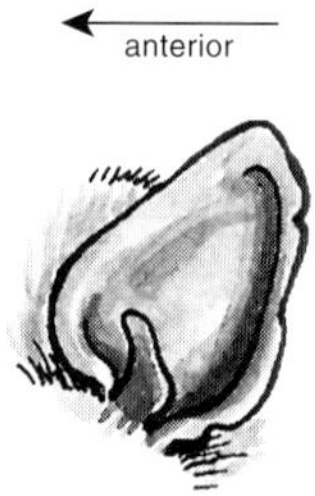

Figure 31 Ear and tragus of Kuhl's pipistrelle

71' Posterior upper incisor extending well above cingulum of anterior incisor; anterior upper premolar a flat-crowned tooth; maxillary toothrow less than 4.5 mm; FA 29–33 mm .... **Anchieta's bat**, *Pipistrellus anchietai*

72. Wing membranes white or translucent; FA 34–38 mm ....**Rendall's serotine**, *Pipistrellus rendalli*

72' Wing membranes dark; FA 29–36 mm 73

73. Dorsal colour paler; wing membranes edged with white; dorsal fur up to 8 mm long; FA 30–32 mm; skull lacking posterior, occipital 'helmet'; only occurring in savanna regions .... **Aloe serotine**, *Pipistrellus zuluensis*

73' Dorsal colour darker; wing membranes not edged with white; dorsal fur not more than 5 mm long; FA 29–36 mm; skull with posterior, occipital 'helmet'; occurring widely in savanna, grassland, montane and semi-arid regions .... **Cape serotine**, *Pipistrellus capensis*

Chapter Six

Species accounts

FAMILY PTEROPODIDAE (FRUIT BATS)

Out of the 14 genera of fruit bats found in Africa and its islands, six occur in southern Africa. There are eight southern African species, of which four are fairly widespread: Wahlberg's epauletted fruit bat (*Epomophorus wahlbergi*), the Gambian epauletted fruit bat (*Epomophorus gambianus*), the straw-coloured flying fox (*Eidolon helvum*), and the Egyptian fruit bat (*Rousettus egyptiacus*). A further four have somewhat marginal or restricted distributions within southern Africa: the Angolan epauletted fruit bat (*Epomophorus angolensis*), Dobson's fruit bat (*Epomops dobsonii*), Bocage's fruit bat (*Lissonycteris angolensis*), and Bergmans's small fruit bat (*Myonycteris relicta*). The Angolan epauletted fruit bat is listed as Lower Risk (Near Threatened), and Bergmans's small fruit bat is listed as Vulnerable, in the *1996 IUCN Red List of Threatened Animals*. The latter species was previously thought to occur only in East Africa, but was recently reported by F.P.D. Cotterill and W. Bergmans to occur in the eastern districts of Zimbabwe.

The Egyptian fruit bat is capable of using a primitive form of echolocation, emitting tongue clicks audible to the human ear. The Gambian epauletted fruit bat (*Epomophorus gambianus*) is taken here to include the subspecies, *E. gambianus crypturus* (Peter's epauletted fruit bat), which some earlier authorities regarded as a separate species. Although fruit bats are typically foliage-roosting, at least three species can be categorised as hollow-roosting, occupying caves or similar cavities: *Lissonycteris angolensis*, *Myonycteris relicta* and *Rousettus egyptiacus*.

Eidolon (Rafinesque 1815)

Straw-coloured flying fox *Eidolon helvum* (Kerr 1792) (Plate 1)

This is by far southern Africa's largest fruit bat, as shown by the measurements. Apart from size (mean wingspan of 750 mm, mean FA of 115 mm, mean mass of 260 g, irrespective of sex), their coloration is distinctive, being an overall orange-yellow above, paler below, with a yellow to orange collar extending from the underside to the sides of the neck. Unlike epauletted bats, they possess a small tail of 15 mm length.

HABITAT

Occurs as a migrant in a variety of habitats.

HABITS

Colonial in their breeding roosts in West, East, and Central Africa, where large colonies of many thousands have been reported. They are unaffected by human activity, roosting in daylight in large cities. A colony of 200 000 individuals is known to roost in a grove of saligna gums (*Eucalyptus saligna*) at Kampala in Uganda.

FEEDING

No data for southern Africa, although, as reported for West Africa, they may feed on the fruit of a range of garden fruits (e.g. plantain, mango, banana, guava, pawpaw, avocada), date palms (e.g. *Borassus*), and the fruits or flower buds of many indigenous species including figs (*Ficus*) and the baobab tree (*Adansonia digitata*).

REPRODUCTION

They do not generally breed in southern Africa, although an apparent breeding colony has been reported from Mozambique (Cotterill, pers. comm.). In West Africa they are reported to give birth to one or occasionally two young.

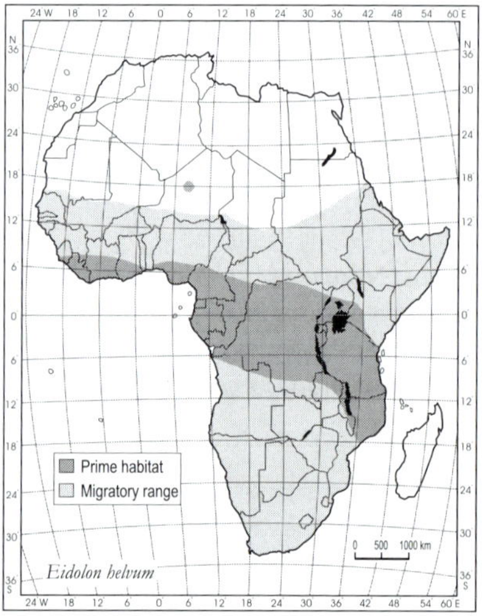

DAY ROOST

Clumps of high trees.

CALL

Reported to chatter continuously throughout the day in their colonies.

DISTRIBUTION AND STATUS

Rare migrant, which occurs sporadically and seasonally throughout the subregion, most records originating from the wetter eastern parts during the summer months.

	Males			Females		
	Mean	Range	*n*	Mean	Range	*n*
FA	115	110–130	44	115	112–133	50
Mass	260	212–311	36	260	183–350	40

Source: ranges and *n* (total distribution) – Bergmans (1990); means irrespective of sex – Skinner and Smithers (1990)

Epauletted fruit bats *Epomophorus* (Bennett 1836)

As the name suggests, all members of this genus possess distinctive long white hairs contained in a glandular pouch in the shoulder region, particularly prominent in males which are capable of erecting these epaulettes during courtship. This feature is also found in the shorter-faced genus, *Epomops,* but not in the remaining genera of fruit bats. Tail is always absent. Sexual dimorphism is pronounced, with males somewhat larger and heavier than females. Three species of *Epomophorus* are currently recognised in the subregion. Unfortunately, size and colour of species overlaps considerably and identification to species is dependent for the moment on the arrangement of transverse ridges on the soft palate.

Angolan epauletted fruit bat *Epomophorus angolensis* (Gray 1870)

Similar in size and colour to the other two species of this genus. Two palatal ridges beyond last molar, as in the Gambian fruit bat, but distinguished from the latter by having fourth palatal ridge closer to 3rd palatal ridge than to 5th, as opposed to halfway between 3rd and 5th. Distribution restricted to southwestern Angola and northwestern Namibia.

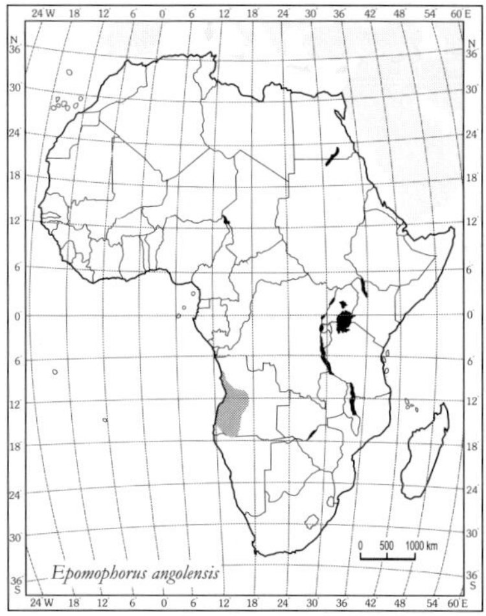

HABITAT, HABITS, FEEDING, REPRODUCTION, AND DAY ROOST

Similar to other epauletted bats, although better adapted to harsh, arid environment where they have been seen roosting in more exposed situations in *Acacia* trees.

DISTRIBUTION AND STATUS

Since it is endemic to a restricted area of Angola and Namibia, it is listed as Lower Risk (Near Threatened) in the *1996 IUCN Red List of Threatened Animals.*

	Males			Females		
	Mean	Range	*n*	Mean	Range	*n*
FA	89	85–94	4	84	81–86	9

Sources: ranges – Bergmans (1988); means – Skinner and Smithers (1990)

Gambian epauletted fruit bat *Epomophorus gambianus* (Ogilby 1835)

(Plate 2)

Note: Includes Peter's epauletted fruit bat (*E. gambianus crypturus*), which some earlier authors considered to be a distinct species. The range of this subspecies encompasses southern Africa.

They are very similar in size and appearance to Wahlberg's epauletted fruit bat with which they may roost in certain parts of the South African lowveld, Zimbabwe and Mozambique; however, the Gambian epauletted fruit bat has two palatal ridges after the last molar, as opposed to one in Wahlberg's epauletted fruit bat.

HABITAT

Evergreen and riverine forests and moist woodlands, with ample cover and fruit-bearing trees.

HABITS

They are gregarious, roosting in groups numbering up to hundreds.

FEEDING

In West Africa they feed on the fruits of garden trees such as bananas and mangoes, and also on the nectar and flowers of a range of trees (which they also pollinate), such as the locust-bean tree (*Parkia*), the sausage tree (*Kigelia* sp.), and baobab (*Adansonia digitata*).

DAY ROOST

Range of evergreen indigenous trees such as figs and sausage trees, which provide a good cover of foliage.

CALL

The monotonous honking male territorial call has been described as a 'musical bark' but is presumably somewhat similar to Wahlberg's fruit bat, although casual observers have reported subtle tonal differences which may prove useful in the future for species identification.

DISTRIBUTION AND STATUS

Occurs widely in Zimbabwe, Mozambique and northern Botswana, but only narrowly in the extreme north and northeast of South Africa, as far south to just north of Port Shepstone on the south coast of KwaZulu-Natal (the only record for KwaZulu-Natal, where it seems uncommon, or has gone unnoticed). Historical records from the Eastern Cape remain unconfirmed and are not included in a recent taxonomic synthesis of the species by Bergmans (1988).

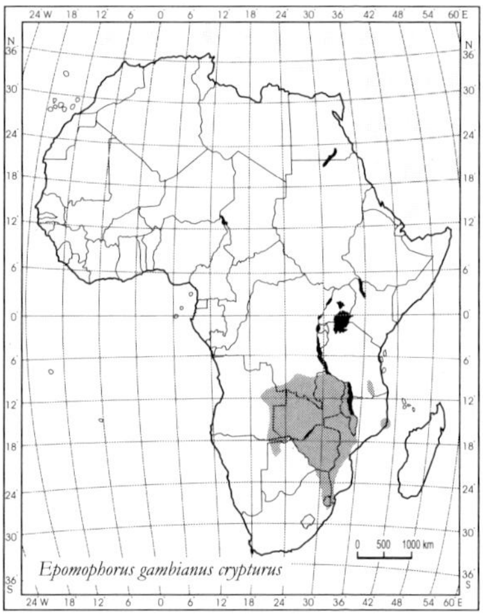

South-central African distribution only

	Males			Females		
	Mean	Range	*n*	Mean	Range	*n*
FA	83	80–88	51	79	75–84	40
Mass	104.5	91–140	11	75.5	56–100	13

Sources: ranges – Bergmans (1988); means – Skinner and Smithers (1990)

Wahlberg's epauletted fruit bat *Epomophorus wahlbergi* (Sundevall 1846)

Light brown to fawn coloration, prominent head, large brown eyes, and conspicuous white hair tufts at base of each ear. Tail absent. Mature males have a white shoulder patch. One ridge on palate behind last molar.

HABITAT

Coastal forest, riverine and Afromontane forest, and bushveld.

HABITS

Gregarious, occurring in colonies varying in size from a few individuals up to 100 or more. In captivity they are intelligent, easily recognise their keepers, and demonstrate individual personalities, as well as an amazing ability to use the claws of the thumb and finger to manipulate food and pull themselves along upside down, or hook themselves up in a semi-horizontal position while releasing the hind claws in order to defecate.

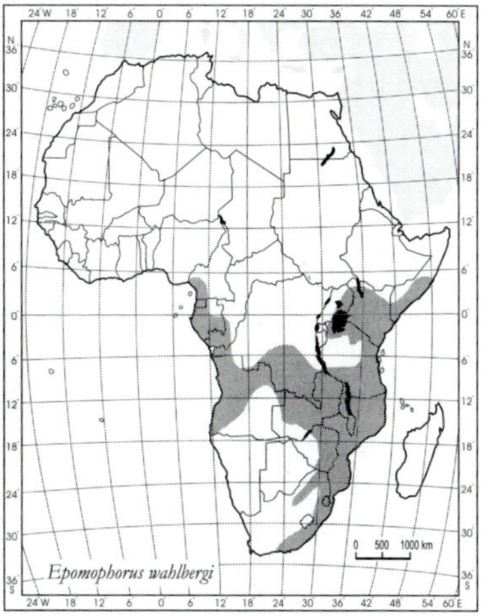

FEEDING

Fruit, nectar, pollen, and flowers. Fruits include a variety of cultivated and indigenous tree species. At Mtunzini in KwaZulu-Natal, the following indigenous trees were recorded from 'spit-outs' collected from under feeding stations (Sapsford, pers. comm.): Cape fig (*Ficus sur*), swamp fig (*F. trichopoda*), Natal fig (*F. natalensis*), wild frangipani (*Voaconga thouarsii*), forest toad tree (*Tabernaemontana ventricosa*), water pear (*Syzigium cordatum*), coastal goldenleaf (*Bridelia mucrantha*), Natal ebony (*Euclea natalensis*), dune myrtle (*Eugenia capensis*), Cape ash (*Ekebergia capensis*), wild custard-apple (*Annona senegalensis*), yellow-wood (*Podocarpus latifolius* and *falcatus*), white milkwood (*Sideroxylon inerme*), quinine tree (*Rauvolfia caffra*), tree fuschia (*Halleria lucida*), marula (*Sclerocarya caffra*), Natal mahogany (*Trichelia emetica*), wild plum (*Harpephyllum caffrum*), red milkwood (*Mimusops caffra*). In Kenya they are known to pollinate baobab flowers. Beetles and other insect remains have occasionally been noted in the stomachs of Wahlberg's epauletted fruit bats. They feed while hovering in front of the fruit or flower, or after landing on a suitable branch.

REPRODUCTION

Gestation is 5–6 months. One (rarely two) young are born at a time, with births occurring throughout the year but with peaks in July and in the summer months. Breeding males display particularly long epaulette hairs as well as a brown discoloration of the skin in the vicinity of the testes (Sapsford, pers. comm.) and in the neck region.

DAY ROOSTS

They prefer tall evergreen trees with good protective cover, including many indigenous trees such as a range of fig species, as well as, in built-up areas, foreign trees such as royal palms and Chinese fan palms (roosting under the fronds along the midrib), casuarinas, and Norfolk pines. In built-up areas in Durban, they have been found roosting under the eaves of houses and under concrete bridges.

Figure 32 Wahlberg's epauletted fruit bat roosting

CALL

Male makes territorial, monotonous, metallic-sounding, honking call to attract females. Groups may emit raucous squabbling sounds like crying human babies during the early hours of the morning.

DISTRIBUTION AND STATUS

Common, even in well-treed urban and suburban areas in the eastern parts of the subregion, as far south as the vicinity of Uitenhage in the Eastern Cape. Owing to their habit of defecating on property walls, and their noisy call, they are regarded to be a nuisance in some suburban areas.

	Males			Females		
	Mean	Range	*n*	Mean	Range	*n*
FA	84	72–95	133	78	68–88	184
Mass	100	60–124	45	95	54–125	78

Sources: ranges and *n* – Bergmans (1988); means – Rautenbach (1982); Skinner and Smithers (1990); Taylor (1998b; unpublished)

Dobson's fruit bat *Epomops* (Gray 1860)

Only one savanna species of this genus occurs in southern and central Africa, while a further two forest-dwelling species are known from equatorial parts of Africa.

Dobson's fruit bat *Epomops dobsonii* (Bocage 1889)

Although related to Epomophorus, *these bats are smaller with shorter faces and more developed yellowish-coloured epaulettes, with hairs up to 18 mm in length compared with 9 mm in* Epomophorus.

HABITAT

Records from outside southern Africa suggest an association with 'wetter Zambesian miombo woodland' (dominated by *Brachystegia, Julbernardia* and *Isoberlina*).

HABITS AND DAY ROOST

Presumably tree-roosting, less gregarious than *Epomophorus*.

DISTRIBUTION AND STATUS

Occur as far north as Rwanda and Tanzania. In southern Africa, known from a single specimen collected along the Chobe River near Kasane in northern Botswana.

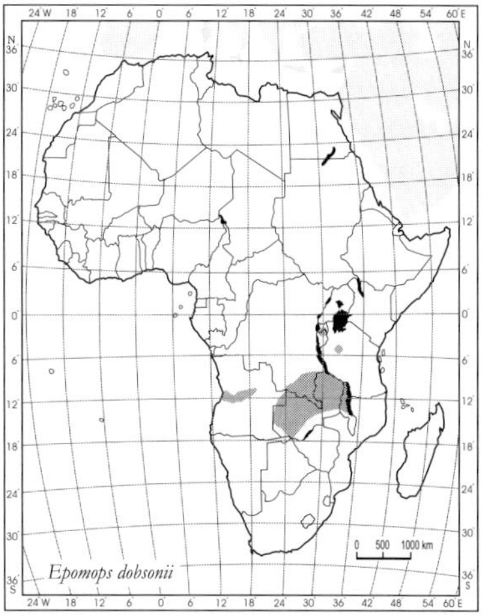

	Males			Females		
	Mean	Range	*n*	Mean	Range	*n*
FA	86	84–92	16	83	80–88	15

Sources (total species range): ranges and *n* – Bergmans (1989); means – Skinner and Smithers (1990)

Bocage's fruit bat *Lissonycteris* (Anderson 1912)

Some previous publications include this genus under *Rousettus.* This genus, which contains only one species, superficially resembles *Rousettus,* but with a shorter face, shorter tail, wing membranes that originate from second (not first) toe and a distinctive glandular throat patch marked by a ruff of long, sticky hairs. These features are shared with the distinctly smaller-sized *Myonycteris* (see Chapter Five). Sexual dimorphism is not marked as in *Epomophorus.*

Bocage's fruit bat *Lissonycteris angolensis* (Bocage 1898)

This species is somewhat smaller (FA 82–90 mm) than the Egyptian fruit bat (FA 83–96 mm), and males possess a distinctive ruff of coarse, sticky hairs. Unlike the Egyptian fruit bat, they cannot echolocate. Compared with the Egyptian fruit bat, they have a much more restricted distribution in southern Africa. Compared with the very similar Bergmans's fruit bat, they are somewhat larger (see Chapter Five).

HABITAT

Similar to Egyptian fruit bat, more dependent on availability of caves than on vegetation type. They have been found in suburban gardens in Mutare, Zimbabwe.

HABITS

They are known to occur in small cave colonies of usually less than 10 (up to 50 recorded in West Africa), often amongst larger numbers of Egyptian fruit bats.

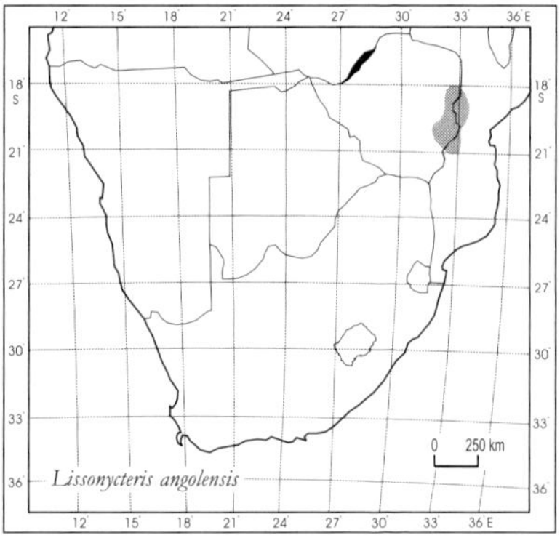

Southern African distribution only

FEEDING

In Mozambique, specimens were netted near a fruiting *Parinari curatellifolia* tree (Cotterill, pers. comm.).

REPRODUCTION

In West Africa, they have been observed to form exclusively female colonies during the breeding season.

DAY ROOST

They are hollow-roosting, occupying caves, where they apparently roost near the entrance, where more light is available (since they cannot echolocate).

DISTRIBUTION AND STATUS

Although also known from Angola, West and East Africa, in southern Africa they are restricted to the eastern parts of Zimbabwe and adjacent parts of Mozambique, extending as far east as Marrameu in the Zambezi Delta.

	Males			Females		
	Mean	Range	*n*	Mean	Range	*n*
FA	85	82–88	3	85	82–90	3

Source: Bergmans (1997)

Bergmans's small fruit bat *Myonycteris* (Matschie 1899)

The genus contains only one southern African species, and has only recently been recorded in the subregion, having previously been recorded from 1 400 km further north.

Bergmans's small fruit bat *Myonycteris relicta* (Bergmans 1980)

This species bears a superficial resemblance to the Egyptian fruit bat but is most closely related to Bocage's fruit bat from which they differ in their smaller size and more restricted distribution.

HABITAT

Recorded from the unique Haroni Forest, south of the Chimanimani Range in eastern Zimbabwe.

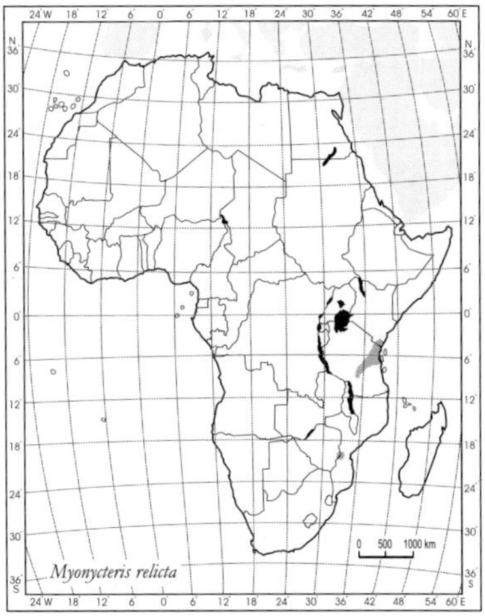

HABITS AND DAY ROOST

Like Bocage's fruit bat, occurring in small colonies in caves.

DISTRIBUTION AND STATUS

Known in southern Africa from a single specimen from the Haroni–Rusitu confluence in eastern Zimbabwe, some 1 400 km south of the nearest recorded locality. Bergmans's small fruit bat is listed as Vulnerable in the *1996 IUCN Red List of Threatened Animals.*

	Males			Females		
	Mean	Range	*n*	Mean	Range	*n*
FA	69	–	1	70	66–75	4
Mass	48	–	1	56	–	1

Source: Bergmans (1997)

Egyptian fruit bat *Rousettus* (Gray 1821)
Four species of this genus are recognised in Africa and its islands, of which *R. egyptiacus* is the most widespread, comprising four subspecies, distributed from India and Arabia to West and East Africa southwards as far as Cape Town. The subspecies occurring in southern Africa is *R. egyptacus leachi*.

Egyptian fruit bat *Rousettus egyptiacus* (E. Geoffroy 1810)
Dark brown on the upper surface and paler below. They are easily distinguished from epauletted bats by the absence of white tufts at the base of the ears, or the white epaulettes of males; also by the presence of a small tail not found in epauletted fruit bats. Unlike epauletted bats they occupy caves during the day. Their smaller size, distinct colour and cave-roosting habits distinguish them from straw-coloured flying foxes. They are distinguished from the genera Lissonycteris *and* Myonycteris *by the absence of a thick ruff of sticky hairs in males and slightly larger size (see Chapter Five). Outside southern Africa, they occur widely in Africa as far north as Egypt and as far west as Senegal. Also known from outside Africa, e.g. Middle East, Arabia and certain Mediterranean countries. The accompanying map shows the distribution of just one subspecies,* R. e. leachi.

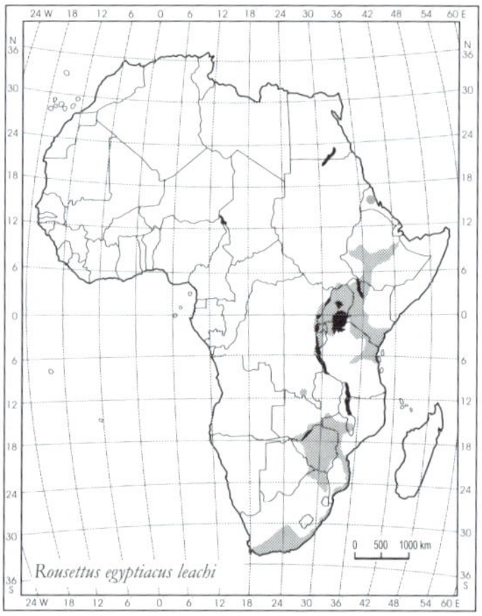

HABITAT
They occur in a wide range of habitats, as long as fruiting trees and caves (or similar hollows) are present.

HABITS
Gregarious and cave-roosting, occurring in cave colonies of up to thousands. They are known to migrate over hundreds of kilometres. The colony occupying Mission Rocks Bat Cave in the Greater St Lucia Wetland Park fluctuates in size

from some 5 000 or more bats in winter to less than 300 in summer, while the opposite pattern occurs in caves in the Tzaneen area of the Northern Province (from 9 000 bats in summer down to some 3 000 in winter), suggesting that the same bats migrate between winter roosts on the KwaZulu-Natal coast and summer breeding roosts in the lowveld. A bat tattooed in the Tzaneen area by N.H.G. Jacobsen was subsequently collected by C. Sapsford at Mtunzini on the north coast of KwaZulu-Natal.

FEEDING

Fruit, nectar, pollen and flowers. Egyptian fruit bats are often regarded as a nuisance during the fruiting season in litchi orchards in the Mpumalanga lowveld. N.H.G. Jacobsen and E. du Plessis showed that there these bats fed on the fruits of other indigenous trees, mainly figs, particularly Cape figs (*Ficus sur*) and *Ficus thonningii*. The bats ate other fruits such as water pears (*Syzygium gerrardii*), water berries (*Syzygium cordatum*), wild plums (*Harpephyllum caffrum*), Cape ash (*Ekebergia capensis*), and bitter almond or red stinkwood (*Prunus africana*). In the Eastern Cape they have been observed feeding on the fruits of Cape ash (*Ekebergia capensis*), saffronwood (*Cassine crocea*), Bushman's poison (*Acokanthera oppositifolia*) and mistletoe (*Viscum obscurum*). In Zimbabwe they have been recorded eating the fruits of *Diospyros senensis*. Elsewhere in Africa, Egyptian fruit bats feed on the pollen and nectar of the baobab (*Adansonia digitata*), sausage tree (*Kigelia africana*) and cotton tree (*Bombax malabaricum*), all of which are pollinated by these bats. In the southern suburbs of Cape Town, Egyptian fruit bats eat the fruits of planted Outeniqua yellow-wood trees (*Podocarpus falcatus*). Since captive Egyptian fruit bats invariably suffer from Vitamin B_{12} deficiency, free-living bats may obtain their requirements from insects ingested incidentally, for example, parasitic wasps which may occur in large numbers inside figs.

REPRODUCTION

After a gestation of four months, a single young (rarely twins) is born. In southern Africa, there is a single birth period per year with peak of births usually in midsummer. The young are carried on the ventral surface of the mother for the first six weeks, and they learn to fly at nine to 10 weeks of age. During the mating season, females and their young form maternity colonies separate from the males.

DAY ROOSTS

Caves, deep crevices and other large cave-like hollows. They often occur in the darkest recesses, although the colony occupying Mission Rocks Bat Cave in northern KwaZulu-Natal is exposed to high levels of light intensity.

CALL

Although they have excellent eyesight, in pitch darkness they use a primitive form of echolocation using repetitive tongue clicks which are audible to humans. Their echolocation call comprises short (1–5 ms), intense, audible, broad-band signals with frequencies between 10 and 60 kHz.

DISTRIBUTION AND STATUS

Common in eastern and northeastern parts of the subcontinent wherever suitable caves are present, occurring coastally as far west as Cape Town, with isolated inland records from the Little Karoo and Karoo. In Cape Town, they are regarded as pests by some people in the southern suburbs when they defecate on house walls and cars. These bats were initially attracted into these suburbs by the palatable fruits of yellow-wood trees (*Podocarpus* spp.) which were planted on road verges by the local authorities.

	Males			Females		
	Mean	Range	*n*	Mean	Range	*n*
FA	89	87–106	51	90	86–101	71
Mass	133	111–166	23	129	100–155	19

Sources: ranges and *n* – Bergmans (1994); means – Skinner and Smithers (1990)

FAMILY EMBALLONURIDAE (SHEATH-TAILED BATS)

The name of this family is derived from the Greek *emballo*, to throw, and *oura*, a tail, referring to the arrangement of the tail which resembles a spear thrown into the ground at an angle. This feature, as well as the large size of their eyes, and their peculiar roosting habits, makes these bats easy to identify to family (see Chapter Five).

The family is represented in the southern African subregion by two genera and three species. While the two species of *Taphozous* are relatively widespread, the sheath-tailed bat (*Coleura afra*) has been recorded only from Tete, on the northern bank of the Zambezi River in Mozambique. As a whole, members of this family can best be categorised as hollow-roosting, preferring cavernous situations such as caves, cellars, or tombs, although in the case of *Taphozous mauritianus*, they prefer the exterior, rather than the interior, walls of houses, and shaded trees.

Sheath-tailed bats *Coleura* (Peters 1867)

Only one species of this genus occurs in southern Africa. A second species is restricted to the Seychelle Islands.

Sheath-tailed bat *Coleura afra* (Peters 1852) (Plate 3)

With a forearm length of 45–55 mm, clearly smaller than other sheath-tailed species. Muzzle pointed, naked and black, with nostrils projecting much further forward than lower lip. Females are slightly larger than males. Colour brown above, paler below, with light, translucent white wings. Large rounded ears and strikingly large eyes. There is no gular sac or radiocarpal pouch.

HABITAT

Open woodland.

HABITS AND DAY ROOST

Hollow-roosting. In East Africa, they prefer caves along the coast and lake shores, where colonies exceeding 50 000 individuals were recorded. In Nigeria they were found in indigenous huts and in cellars. The type-specimen was taken from a colony in a dark cellar at Tete, Mozambique. Colonies break up into harem clusters dominated by males.

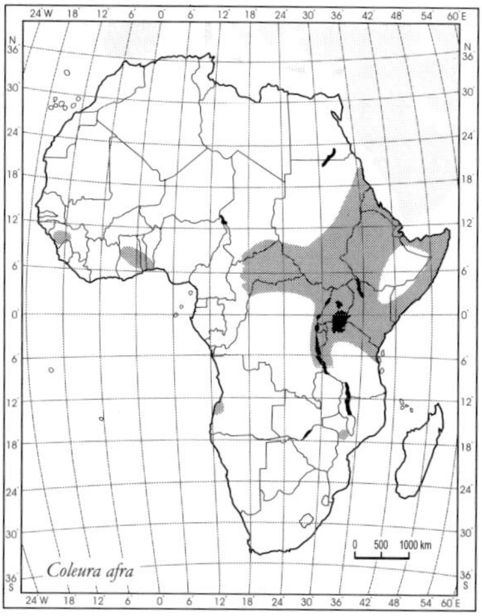

REPRODUCTION

In Kenya, females give birth to a single young twice a year, during both the short rainy season and the long rainy season. Gestation is 16 weeks during the long rains (March–June) and 15 weeks during the short rains (November–December).

DISTRIBUTION AND STATUS

They occur as far north as West Africa to southwestern Arabia. In southern Africa, known only from one specimen collected from the Zambezi River, in the Tete District of Mozambique.

	Males			Females		
	Mean	Range	*n*	Mean	Range	*n*
FA	46	(45–55)	–	49	(45–55)	–
Mass	–	(10–12)	–	–	(10–12)	–

Sources (total distribution): means – McWilliam (1987); ranges irrespective of sex – Nowak and Paradiso (1983)

Tomb bats *Taphozous* (E. Geoffroy 1818)

These are among the largest of the insectivorous microbats, making them easy to recognise. Their habit of roosting on the vertical surface of walls and tree trunks face-down, and moving rapidly sideways in a crab-like fashion when disturbed, is characteristic of these bats.

Mauritian tomb bat *Taphozous mauritianus* (E. Geoffroy 1818)

One of the largest and most attractive of the insect-eating bats, with prominent eyes and relatively long snout (Figure 33). Greyish-grizzled upper surface and pure white undersurface; colour distinguishes it from the Egyptian tomb bat. Males have a small dark circular glandular pouch situated under the chin (see colour illustration). A radiocarpal pouch is present on the undersurface of the wings (Figure 34); naked wing membranes.

HABITAT

Open savanna and forest edges in the wetter northern and eastern parts of the subcontinent.

HABITS

Groups of two or more, up to about 30, usually much less, comprising one or more females and their young. Males occur solitarily in the vicinity of females, except during mating. Mating is brief, noisy and violent, with males biting the necks of females, and females protesting with loud squeals.

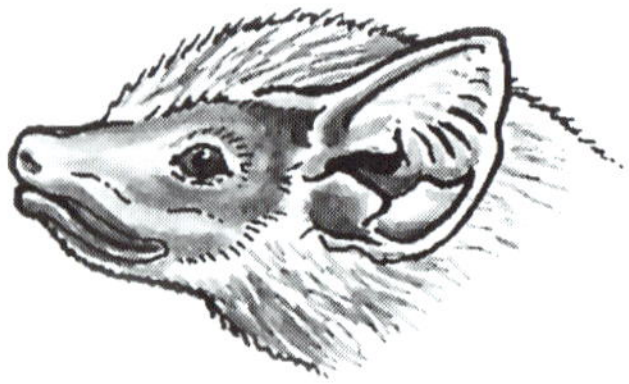

Figure 33 Head of Mauritian tomb bat

Figure 34 Wing showing wing (radio-carpal) pouch of Mauritian tomb bat

FEEDING

They are moth specialists, with termites and butterflies occurring in their diet at lower frequency. The presence of butterflies in the diet suggests that they may occasionally forage during daylight, aided by their excellent eyesight. As suggested by their long, narrow wings, they are fast-flying aerial feeders, feeding in open spaces.

REPRODUCTION

Young of various sizes have been recorded throughout the year in Zaire, where the gestation has been recorded as four months, while in Malawi and KwaZulu-Natal, two distinct birth periods have been described: November–December and March–April in Malawi; and November–December and February–March in KwaZulu-Natal. A single young is born at a time.

DAY ROOST

They do not roost in tombs as implied by their name, roosting instead on the outer bark of trees, under covering vegetation. In built-up areas they are found on tall trees such as Chinese fan palms and royal palms, under overhanging vegetation, or, more frequently under the eaves on the outer walls of houses, particularly face-brick houses. They tolerate high levels of daylight while roosting, as well as close human presence, but appear to be constantly vigilant, with eyes open, during the day.

CALL

Short-duration (2–4 ms), multi-harmonic (up to three harmonics), steep-FM call, with very low (audible) fundamental frequency (10–16 kHz; loudest at 13 kHz). It is likely that open-foraging moth specialists like this species use low echolocation frequencies because these fall below the sensitivity range (typically 20–50 kHz) of hearing moths, preventing them from detecting the hunting calls of the bats and taking early avoidance measures.

DISTRIBUTION AND STATUS

It occurs widely throughout Africa south of the Sahara, but generally known from few collecting records. In southern Africa it occurs mostly in the northeastern and eastern parts, extending narrowly along the east coast of South Africa to as far south as around Mossel Bay, with an isolated inland record from Hartwater in the Northern Cape. Apparently it is more common than collecting records indicate, as, with concentrated searching by bat-workers, over 20 roosts have been identified in the Durban area.

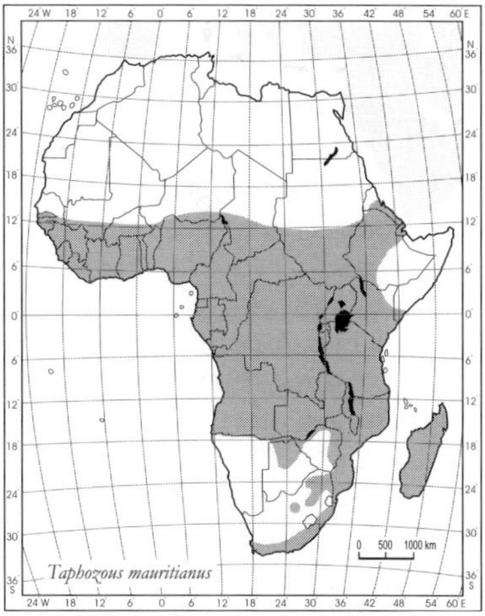

	Males			Females		
	Mean	Range	*n*	Mean	Range	*n*
FA	62	60–64	22	62	61–66	11
Mass	28	20–36	17	29	27–32	10

Sources: Rautenbach (1982); Skinner and Smithers (1990); Taylor (1998b)

Egyptian tomb bat *Taphozous perforatus* (E. Geoffroy 1818)

Similar in size but best distinguished from the Mauritian tomb bat in having a uniform dark brown upper surface (as opposed to grizzled greyish), and the under parts not white. Males have no gular sac (present in the previous species). They have very large eyes.

HABITAT

Open woodland near rocks or stone buildings where they roost.

HABITS

Colonies of around six to eight are commonly encountered, but may number hundreds in southern Zimbabwe, in sandstone overhangs (Cotterill, pers. comm.).

DAY ROOST

Originally described from the royal tombs of Egypt. They roost in rocky outcrops and stone or brick buildings, but, unlike the previous species which roosts in the open, they roost tucked away in darkened narrow crevices in rocks or brickwork. Unlike the more solitary Mauritian tomb bat, Egyptian tomb bats roost in larger groups of hundreds, with individuals bunched together.

FEEDING

The diet comprises predominantly, in order of priority, moths, termites, beetles, and, to a lesser extent, crickets and katydids, bugs, lacewings, ants, and flies. Like the previous species, they are high aerial feeders.

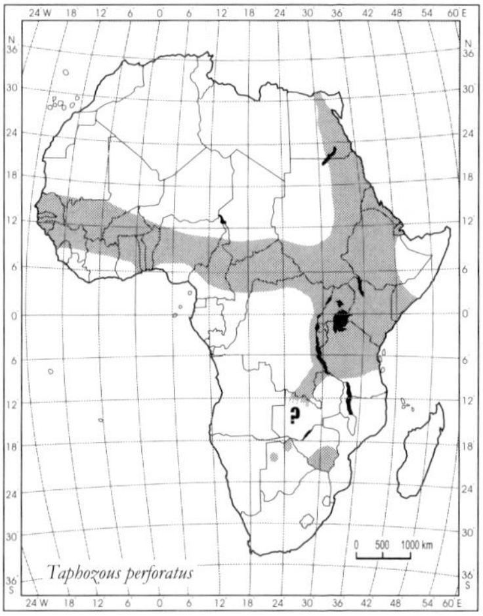

DISTRIBUTION AND STATUS

Occur widely from northwest India to much of Africa; however, in southern Africa, they are restricted in distribution to southern Zimbabwe and in the Okavango Delta and Shashi–Limpopo confluence areas of northern and eastern Botswana respectively.

	Males			Females		
	Mean	Range	*n*	Mean	Range	*n*
FA	–	–	–	62	62–63	8
Mass	–	–	–	30	23–33	8

Source: Skinner and Smithers (1990)

FAMILY MOLOSSIDAE (FREE-TAILED BATS)

Members of this family are easily identified by the mouse-like tail, which extends beyond the hind margin of the tail membrane. Free-tailed bats are sometimes referred to as mastiff bats or bulldog bats, referring to their thick, wrinkled lips, and the family name is derived from the Greek *molossus* which refers to a kind of dog used by Greek shepherds in ancient times. Molossids are generally gregarious and colonial. They all have narrow pointed wings, adapted for fast flight in open spaces well above the vegetation, and all can scurry on all fours horizontally, forwards, backwards and sideways in the narrow spaces where they typically roost during the day. Generally members of the family can be categorised as crevice-roosters, many of them taking readily to suitable crevices in the roof spaces of houses and other buildings, and the expansion joints of bridges.

Free-tailed bats can often be identified whilst in flight using an inexpensive heterodyne (tunable) or frequency division bat detector (see Chapter Eight), as they emit loud, low frequency, 'narrow-band' or 'shallow' FM clicks with a distinctive harsh tonal quality. Their flight behaviour is also somewhat distinctive as they forage high up in the air above the vegetation, flying fast at heights of at least 20 m above the ground, making occasional rapid forays downwards to catch lower-flying insects. This is quite distinct from the behaviour of plain-faced bats which often dart, flutter and jink just a few metres above the ground.

Fourteen molossid species and four genera are found in southern Africa. The molossid group has adapted to living in warm tropical climates, being largely absent from colder temperate climates. Species richness declines southwards in southern Africa. Thus, of the 14 species found in the subregion, only four are commonly encountered in South Africa, *Mormopterus petrophilus, Tadarida aegyptiaca, Mops condylurus* and *Chaerephon pumilus*. *Mops midas* and *Tadarida fulminans* are known only from the Kruger National Park and extreme northern parts of Mpumalanga, while a further two species are represented by only one or two specimens in South Africa, *Tadarida ventralis* and *Mormopterus acetabulosus*.

Three genera, *Tadarida*, *Chaerephon* and *Mops* are not very clearly demarcated, since characteristics on which they are based can be variable, and have in the past been included as subgenera of one genus, *Tadarida*. Problems can sometimes be encountered in identifying medium-sized species of these three genera in the field and in assigning individuals to a particular genus. Thus, for example, *Chaerephon ansorgei* does not clearly show the post-aural crest of males typical of *Chaerephon*, and has ears barely joined, as well as a marked palatal emargination, typical of *Tadarida*. The distinctive blackish 'beard' found in this species is occasionally absent or unclear. Similarly, *Chaerephon bivittatus* has a marked palatal emargination, and moreover sometimes does not show the distinctive white markings typical of this species. Even within one species, *Chaerephon pumilus*, the degree of palatal emargination varies considerably. Despite these problems, the Key in Chapter Five should be effective in almost all practical instances, particularly taking into account distributions. Where practical, specimens should be lodged with a natural history museum for careful analysis of the prepared skin and skull.

Wrinkle-lipped bats *Chaerephon* (Dobson 1874)

The generic name *Chaerephon* is derived from the friend of Socrates who is depicted as 'Chaerephon the bat' in Aristophane's comedy Aves. Members of this genus can be distinguished by the typical presence of a crest of hairs behind the forehead of males (postaural crest), which usually lie back against the neck but can be erected at will. A white or off-white band of hairs is typically present, running longitudinally between the flanks and wing membranes. Like *Mops*, but not *Tadarida*, the ears are joined across the top of the head by a band of skin. While this genus includes the two smallest southern African free-tailed bat species (FA <43 mm), the pale free-tailed bat and the little free-tailed bat, other members are larger (FA 43–50 mm), but these larger species tend to have distinctive white head or body markings or a conspicuous blackish beard, which help to distinguish them from species of *Mops* and *Tadarida* (see Chapter Five).

The genus *Chaerephon* comprises five southern African species: Ansorge's free-tailed bat (*C. ansorgei*), the spotted free-tailed bat (*C. bivittatus*), the pale free-tailed bat (*C. chapini*), the Nigerian free-tailed bat (*C. nigeriae*), and the little free-tailed bat (*C. pumilus*). The Nigerian free-tailed bat and pale free-tailed bat are more sparsely distributed and less frequently collected than the remaining species.

Ansorge's free-tailed bat *Chaerephon ansorgei* (Thomas 1913)

Very similar in colour to the Egyptian free-tailed bat, umber-brown above, underparts a shade paler, with throat having longer and darker, almost black hairs (this blackish 'beard' is usually but not always present on both sexes); a reddish colour phase, with darker red beard, has been described in some individuals of this species; wing membranes are translucent light brown; FA 43–48 mm, larger than the little free-tailed bat (FA 35–40 mm), overlapping with the Angolan free-tailed bat (FA 43–49 mm), but without the white underparts and naked band across the shoulders. It lacks the lateral white band of hair running along the inner edge of the wing, and has the ears barely joined together, as well as poorly developed or absent post-aural crests in males; these last three features, in addition to its wide palatal emargination, are features more typical of Tadarida *than* Chaerephon, *suggesting that it should perhaps be better referred to* Tadarida.

HABITAT

Dry woodland savanna in the Kruger National Park (*Adansonia digitata-Colophospermum mopane*-dominated savanna woodland) and elsewhere.

HABITS

Little known; presumably similar to little free-tailed bat.

FEEDING

Insects taken include beetles, lacewings, and moths. They are fast-flying, high aerial feeders.

DAY ROOST

Crevice-roosting.

CALL

Long duration (15 ms), multi-harmonic (up to two), narrow bandwidth (16–28 kHz) shallow-FM calls, with low dominant frequency of 18 kHz.

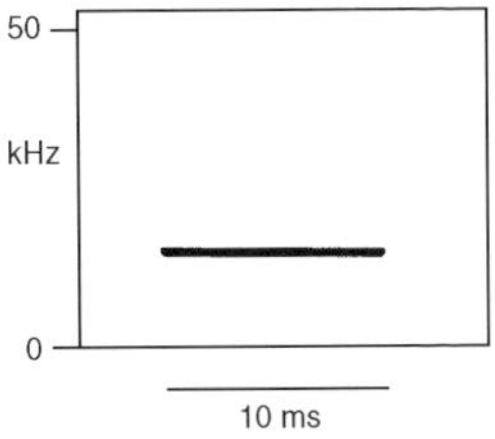

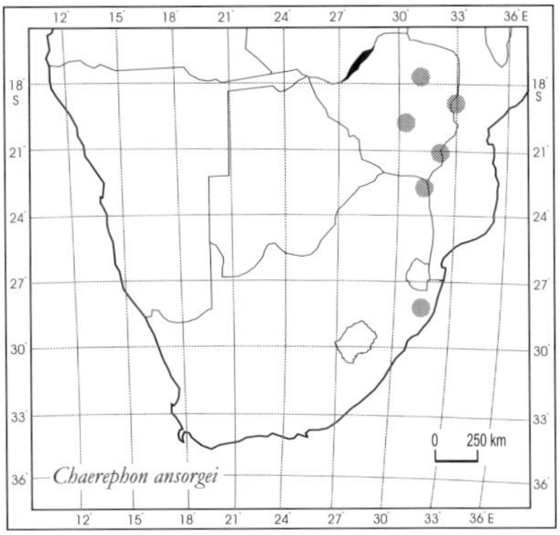

Southern African distribution only

DISTRIBUTION AND STATUS

Widespread in Zimbabwe, and locally very common near suitable roosts. Known from just two records from South Africa, the Kruger National Park and Mkuzi Game Reserve. Listed as Specially Protected in the new KwaZulu-Natal Conservation Ordinance of 1999.

	Males			Females		
	Mean	Range	*n*	Mean	Range	*n*
FA	45	43–47	61	45	43–47	58
Mass	15	11–18	61	14	11–18	58

Source (Democratic Republic of Congo): Taylor (1999b)
Note: for South Africa FA 46–48, irrespective of sex, *n* = 9 (Rautenbach et al. 1985)

Spotted free-tailed bat *Chaerephon bivittatus* (Heuglin 1861)

Dark umber-brown upper parts with slightly paler underparts. As the name suggests, the head, shoulders and possibly flanks are variably speckled with white flecks or short stripes. Wings and interfemoral membrane are dark brown. Apart from its distinctive colour pattern, its larger size (FA 43–47 mm) easily distinguishes it from the smaller two species of Chaerephon *(FA 40 mm or less),* C. pumila *and* C. chapini. *The absence of a blackish beard also distinguishes this species from the similar-sized* C. ansorgei.

HABITAT

Brachystegia woodland in Zimbabwe.

HABITS

Found in small groups of up to six.

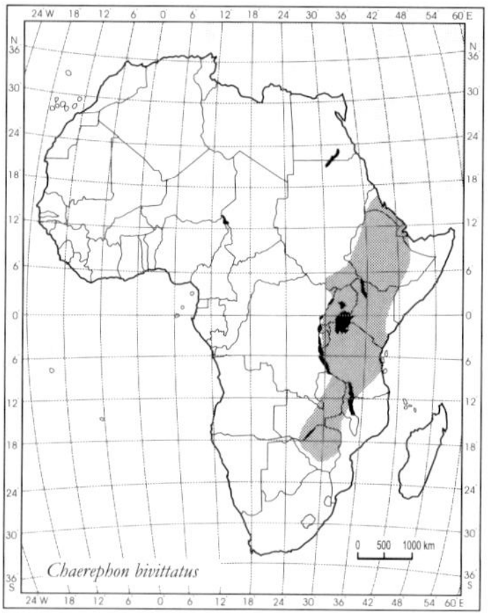

DAY ROOST

Rock crevices, hollows in trees, and under the loose bark of trees.

DISTRIBUTION AND STATUS

Occurring widely in Africa from Eritrea, Ethiopia and Sudan in the north to as far south as Bulawayo in Zimbabwe.

	Males			Females		
	Mean	Range	*n*	Mean	Range	*n*
FA	47	45–47	10	46	43–47	21
Mass	15.4	14–17	10	14.6	13–17	21

Source: Skinner and Smithers (1990)

Pale free-tailed bat *Chaerephon chapini* (J.A. Allen 1917)

Easily distinguished by the pale fur colour, pale cinnamon-brown above, flecked with white, and greyish-brown below with a whitish band reaching from throat to sternum. Males have a distinctive crest of long, erectile, bicoloured hairs (whitish below, reddish towards tip) behind the forehead. Wing membranes are white, tinged with yellow. A white longitudinal band is present between the wing membrane and flanks. FA 34–39 mm.

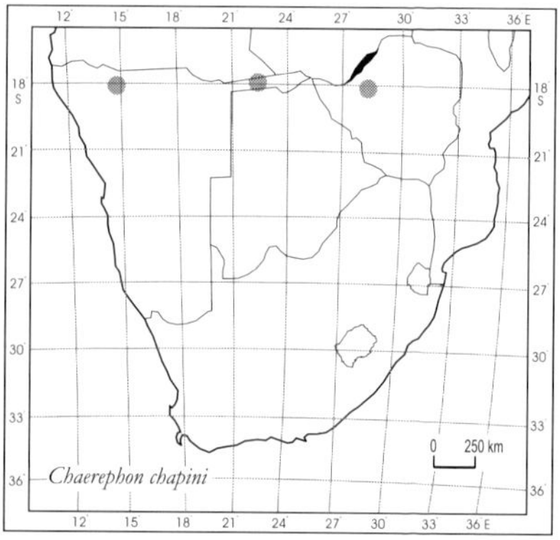

Southern African distribution only

HABITAT
Dry woodland.

HABITS AND DAY ROOST
Unknown in southern Africa, but undoubtedly crevice-roosting (and possibly roof-dwelling) like other members of the family.

FEEDING
Predominantly beetles and moths, but also flies. They are high aerial feeders.

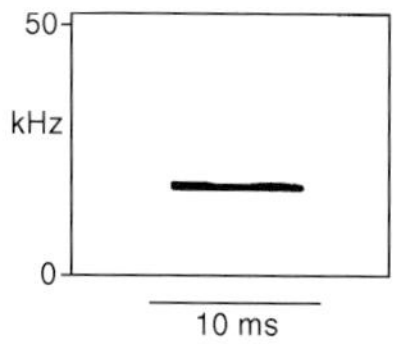

CALL
Long duration (10 ms), narrow-band (19–27 kHz) FM call with low dominant frequency of 20 kHz.

DISTRIBUTION AND STATUS
It is known from scattered localities in Central and East Africa, occurring marginally in the north of Zimbabwe, Botswana and Namibia. The species is listed as Lower Risk (Near Threatened) in the *1996 IUCN Red List of Threatened Animals.*

Nigerian free-tailed bat *Chaerephon nigeriae* (Thomas 1913)

Forearm varies from 44–50 mm. Colour very dark brown above and below, contrasting with white flank hairs which extend onto wing membrane. The wing and interfemoral membranes are off-white. An erectile crest of hairs is present behind the forehead in males. Tragus symmetrically bilobed in this species, but asymmetrically bilobed in the little free-tailed bat. FA 44–50 mm.

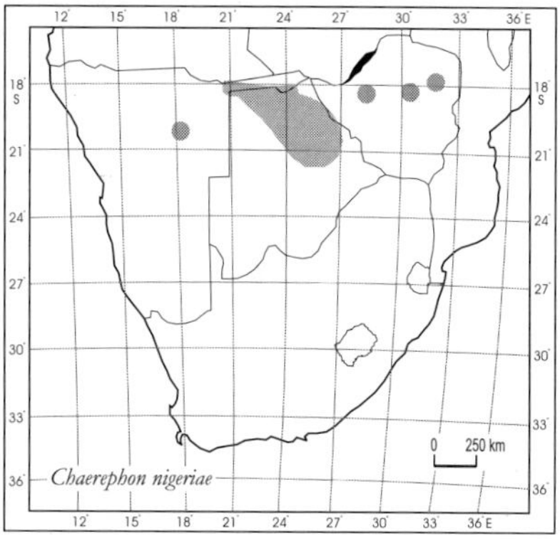

Southern African distribution only

HABITAT
Brachystegia woodland in Zimbabwe.

HABITS
Gregarious, as small groups have been collected together.

DAY ROOST
Roofs, thatched huts, and under the bark of dead trees. In West Africa, specimens were taken from the roofs of brick and thatched houses, as well as from eaves.

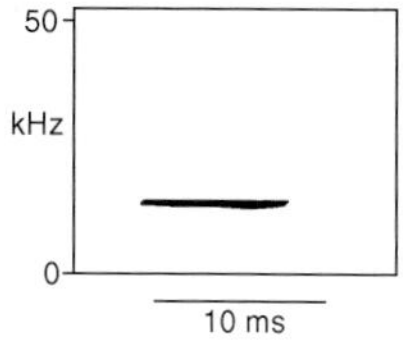

CALL
Long-duration (10 ms), narrow-band (16–26 kHz) FM call with low dominant frequency of 17 kHz.

DISTRIBUTION AND STATUS

Wide but scattered distribution throughout most of tropical Africa, from Niger and Ethiopia, southwards to northern Botswana, Namibia and Zimbabwe.

Little free-tailed bat *Chaerephon pumilus* (Cretzschmar 1826) (Plate 4)

Along with the pale free-tailed bat they are the smallest of the free-tailed bats, FA 35–40 mm. The postaural crest of males is conspicuous, especially when the skin between the ears is gently stroked forwards to raise the stiffened hairs of the crest which usually lie back against the head. Colour blackish-brown to brown or greyish-brown above, slightly paler below; dark winged and light winged forms are found, with light winged forms tending to have a narrow, mid-ventral paler band from anus to chest. A narrow band of white to tawny coloured hair runs along the inner edge of the wing membrane from forearm to thigh; this is indistinct in some individuals.

HABITAT

They occupy a wide range of habitats, including savanna, mountainous and arid areas, and are very common in dwellings in built-up areas within their distribution. Observations in Zimbabwe, Swaziland and KwaZulu-Natal show their preference for lowveld and coastal areas, rarely occurring at altitudes in excess of 1 000 m.

HABITS

Gregarious; colony sizes of around 10 to up to an estimated 2 500 have been recorded in buildings and holiday resort complexes in KwaZulu-Natal.

FEEDING

In order of priority, their diet comprises bugs, beetles, moths, Hymenoptera, and to a lesser extent, nocturnal flies. They are fast, high-flying, aerial feeders.

REPRODUCTION

After a 60 day gestation, a single young is born. There are three births per year (early November, late January, and early April), a fact attributed to their ability to undergo a post-partum oestrus. Females become sexually mature at the age of 5–12 months.

DAY ROOST

Crevices in trees, rocks or, very commonly, in the roofs of houses. In roofs, their preferred roosting sites include the horizontal surfaces of rafters, between the rafters and brickwork, in the corrugations of corrugated iron roofs, between the roof tiles and the plastic lining underneath, between the thatch and timber in thatched roofs, as well as in broken air vents.

CALL

Long duration (16 ms), multi-harmonic (up to two), narrow bandwidth (21–33 kHz) shallow-FM calls with a low dominant frequency of 23–28 kHz.

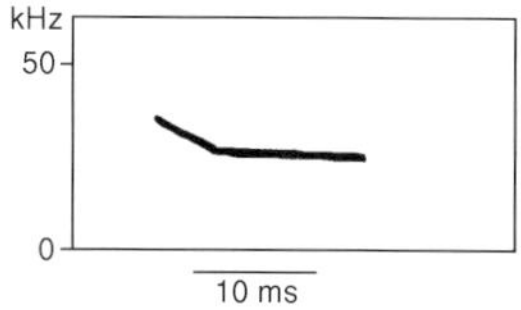

DISTRIBUTION AND STATUS

Occurring very widely in sub-Saharan Africa north of southern Africa, but, in the subregion, occurring in northern and eastern parts. Very common in built-up areas and safari lodges in the lowveld of Swaziland and northeastern South Africa, where they are frequently regarded to be a nuisance because of the accumulation of guano and the smell and chittering noises associated with large colonies. Until recently, populations of the flat-headed free-tailed bat occurring in the Western Cape were mistakenly presumed to belong to this species (Jacobs and Fenton, In press).

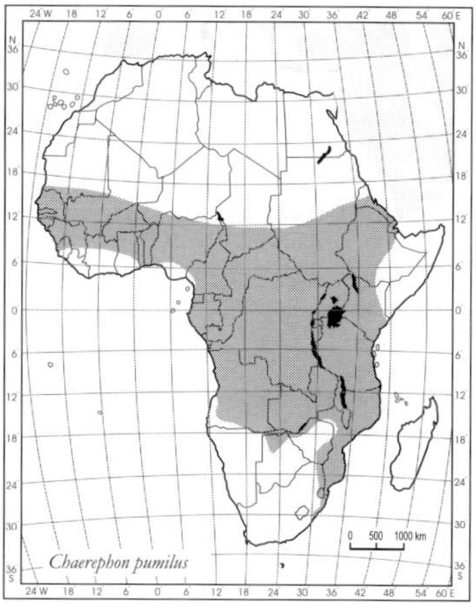

	Males			Females		
	Mean	Range	*n*	Mean	Range	*n*
FA	37	36–40	36	37	35–39	55
Mass	11	10–14	37	12	10–16	81

Sources: Smithers (1971); Smithers and Wilson (1979); Taylor (1998b; unpublished)

Mops free-tailed bats *Mops* (Lesson 1842)

The best character for recognising members of this genus, namely, the reduction of the third molar to a V-shaped cusp, as opposed to the full N-shape, is not much help in the field. However, the presence of a paler band of near-naked skin across the shoulders, the huge size of the Midas free-tailed bat (FA 59–61 mm), and the predominantly white belly of the only other species, the Angolan free-tailed bat, should make identification of the two southern African species reasonably easy. The genus also has a relatively conspicuous antitragus (Figure 35).

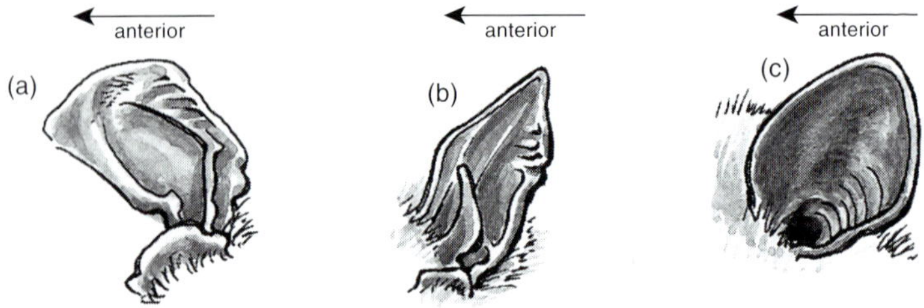

Figure 35 Ears of different bats to illustrate the presence or absence of tragus and antitragus.
(a) Antitragus large relative to tragus of the Midas free-tailed bat
(b) Smaller antitragus relative to tragus of the yellow house bat
(c) No antitragus or tragus of the straw-coloured fruit bat

Angolan free-tailed bat *Mops condylurus* (A. Smith 1833) (Plate 5)

Smaller than the Midas free-tailed bat, colour is dark brown above with paler, sparse-haired band across shoulders, much paler below, with a variable band of white extending mid-ventrally between the throat and the anus and onto the thighs. The ears are joined by a substantial band of skin, as found in Chaerephon *but not* Tadarida. *The naked skin of the wing membranes extends onto the body dorsally, forming a clear line of demarcation with the lateral edges of the body fur.*

HABITAT

Found in a wide range of vegetation associations.

HABITS

Gregarious, occurring in colonies varying in size from fewer than 100 to over 1 000, although usually numbering up to dozens rather than hundreds.

FEEDING

Probably beetles, based on the powerful jaws of this species. They are high aerial feeders.

REPRODUCTION

After a gestation of 85 days, they give birth to a single young, with two pregnancies per year, between September and early May.

DAY ROOST

Caves, rocky crevices, and, commonly, in crevice-like spaces in roofs, often found roosting together with colonies of the little free-tailed bat. In West Africa they occur naturally in tree hollows rather than in rocky situations, but are also a nuisance in the roofs of houses.

CALL

Medium duration (7 ms), steep-FM calls with a low dominant frequency of 24 kHz (range 22–39 kHz) have been recorded from individuals just prior to landing.

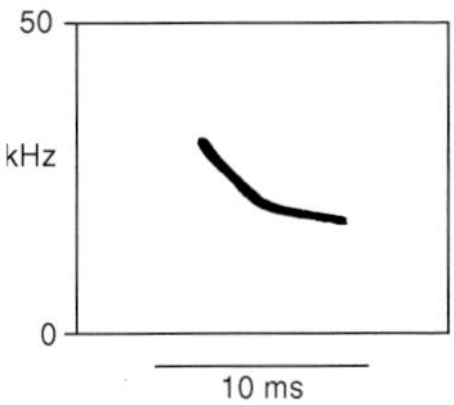

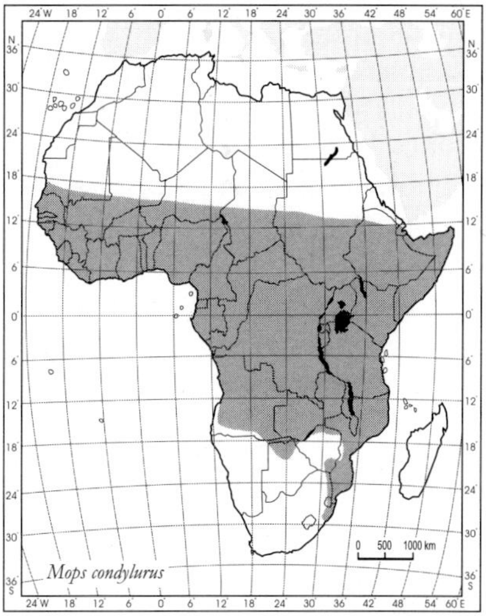

DISTRIBUTION AND STATUS

Distribution is virtually identical to that described for the little free-tailed bat with which it often co-occurs; however, it does not occur as far southwards along the KwaZulu-Natal coast, and is not found in the Durban area where the little free-tailed bat is very common. Common, often regarded as a pest in buildings due to the pungent smell and large quantities of guano produced by large colonies. Humane exclusion methods coupled with the use of specially designed bat houses to accommodate evicted colonies has been successful with this species, and several occupied bat houses occur in the lowveld of Mpumalanga, KwaZulu-Natal and in Mozambique.

	Males			Females		
	Mean	Range	*n*	Mean	Range	*n*
FA	47	43–49	15	47	45–48	12
Mass	22	17–28	15	20	16–28	12

Source: Skinner and Skinners (1990)

Midas free-tailed bat *Mops midas* (Sundevall 1843)

Dark brown above, paler tawny with white-tipped hairs below. Like the Angolan free-tailed bat, they have a band of sparse hairs across the shoulders. Next to the large-eared free-tailed bat, it is the largest free-tailed bat, FA 59–61 mm.

HABITAT

Associated with major rivers in savanna woodland.

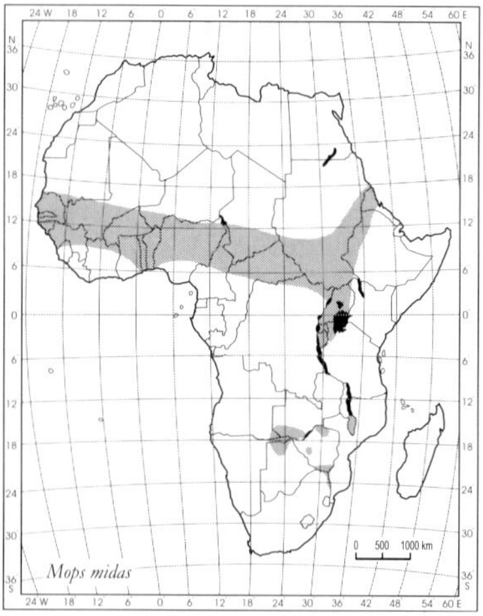

HABITS

Gregarious, occurring in colonies of up to hundreds.

DAY ROOST

Recorded from a dark attic and from expansion joints in a concrete bridge. In West Africa they occupy crevices in taller trees where different colonies may occupy different openings in the same tree.

CALL
Shallow FM call with a low maximum frequency of 30 kHz, and duration of 12 ms.

DISTRIBUTION AND STATUS
In South Africa, only likely to be encountered in the Kruger National Park and the extreme northern parts of Mpumalanga Province; otherwise known from southern, western and northern Zimbabwe and northern Botswana.

	Males			Females		
	Mean	Range	*n*	Mean	Range	*n*
FA	61	59–61	3	60	59–61	4
Mass	49	42–52	4	45	41–48	4

Source: Skinner and Skinners (1990)

Flat-headed bats *Mormopterus* (Peters 1865)

The genus *Mormopterus* contains two southern African species, the flat-headed free-tailed bat (*M. petrophilus*), and the Natal free-tailed bat (*M. acetabulosus*). Members of the genus can easily be distinguished from other free-tailed bats by having ears which are clearly separated.

Natal free-tailed bat *Mormopterus acetabulosus* (Hermann 1804)

The Natal free-tailed bat (*M. acetabulosus*) occurs on the Indian Ocean islands of Madagascar, Mauritius, and Reunion, and was described from southern Africa from a single specimen collected in Durban in 1833 by Andrew Smith, and a further undated specimen from Durban in the National Museum of Natural History in Paris. It has only been recorded from one other locality on the African mainland, in Ethiopia. It is listed as Vulnerable in the *1996 IUCN Red List of Threatened Animals* and is probably not continually resident in southern Africa. However, there is always the possibility of locating a further specimen of this bat on the southern African mainland.

	Males		
	Mean	Range	*n*
FA	40	–	1

Source: National Museum of Natural History (MNHN) (Paris) collection

Flat-headed free-tailed bat *Mormopterus petrophilus* (Roberts 1917)

Upper parts pale to dark brownish; underparts white to greyish-white. As the name suggests the skull is flattened dorso-ventrally, allowing these bats to retreat into very narrow rocky crevices. Ears pointed and separate; warty muzzle; main brush of hairs above nose. In the drier western parts of the South Africa, Botswana, and Namibia, this species co-occurs with only one other free-tailed bat, the Egyptian free-tailed bat, from which it is easily identified by its smaller size; however, it co-occurs with more species in northern South Africa and Zimbabwe.

HABITAT
Rocky areas where crevices are available.

HABITS
Small groups of up to four recorded, clustered tightly together.

DAY ROOST
Under exfoliated granite sheets or in narrow rocky crevices.

DISTRIBUTION AND STATUS
Scattered distribution in southeastern Zimbabwe, Mozambique, Namibia, Botswana and drier western and northern parts of South Africa. Populations from the Western Cape previously assigned to *Chaerephon pumilus* (Skinner and Smithers 1990) have recently been referred to this species (Jacobs and Fenton, In press).

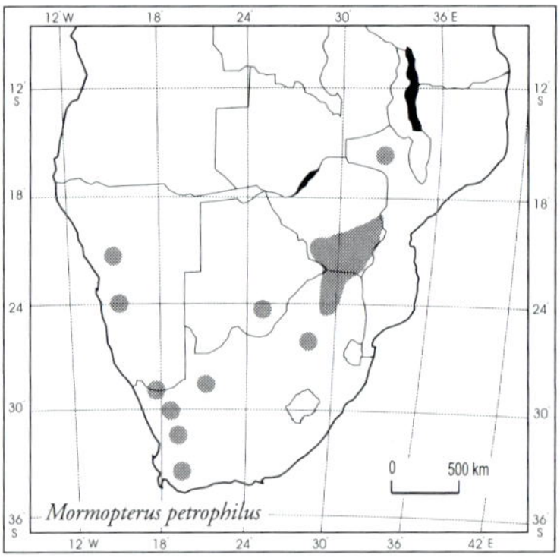

	Males			Females		
	Mean	Range	*n*	Mean	Range	*n*
FA	42	38–48	8	44	41–51	7
Mass	13	9–16	9	15	13–22	6

Source: Rautenbach (1982)

Mastiff bats *Otomops* (Thomas 1913) (Plate 6)

Large-eared free-tailed bat *Otomops martiensseni* (Matschie 1897)
Large size, conspicuously large, round-pointed, forward-projecting ears, and protruding, pig-like snout make this species unmistakable; also paler collar extending across top of shoulders. Males are larger than females.

HABITAT
Apart from two isolated records from savanna woodland in Zimbabwe, in South Africa they are restricted to the roofs of (mostly older, multi-storey) houses and apartments in built up areas within greater Durban and the South Coast.

HABITS
Small groups typically comprise a single dominant male with a harem comprising one or a few females with their young. Larger groups of up to 30 or more have been found, as well as single males.

FEEDING
Moths make up 97% of their diet based on faecal pellet analyses. They are fast-flying, high aerial feeders.

REPRODUCTION
After a three-month gestation, females give birth to a single young. In East Africa, births occur between October and January, and in May and June. In Durban, juveniles have been recorded between October and May, but it is not known whether individual females can give birth more than once a year.

DAY ROOST
In Durban, they roost inside the roof spaces of houses, usually on the inside gable apex wall, or on the vertical surface of rafters close to the eaves. In Kenya, and presumably elsewhere in Africa where they occur, they roost in large colonies in caves.

CALL
In Durban, very long duration (maximum 60 ms), very shallow ('quasi-constant frequency', QCF), low frequency (24–28 kHz) search phase calls have been recorded, as well as long (28 ms), audible (7–16 kHz, loudest at 11 kHz), multi-harmonic calls emitted in pairs on emergence from their day roost. The latter call is most similar to that recorded by M.B. Fenton at Sengwa in Zimbabwe, who termed this species 'the audible bat', owing to its extremely low echolocation frequency. In open-foraging moth spe-

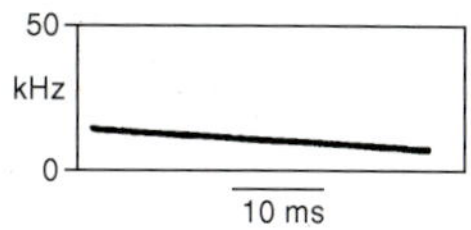

cialists like this species, low frequency echolocation calls are probably an adaptation to avoid the range of maximum frequency sensitivity of hearing moths, which is usually in the region of 20–50 kHz.

DISTRIBUTION AND STATUS

Within southern Africa, known only from two isolated records from Zimbabwe, as well as within the greater Durban area, with isolated records extending northwards along the coast to Ballito, southwards to Margate, and inland to Pietermaritzburg. Outside southern Africa, scattered records are known from East Africa and northwards to Yemen (Harrison, pers. comm.). The Durban population is recognised to be a distinct subspecies, *Otomys martiensseni icarus*. The species is listed as Vulnerable in the *1996 IUCN Red List of Threatened Animals*. Since October, 1996, listed in the Endangered Mammal category under the 1974 Natal Conservation Ordinance (Schedule 6), and listed as Specially Protected in the revised KwaZulu-Natal Conservation Ordinance of 1999.

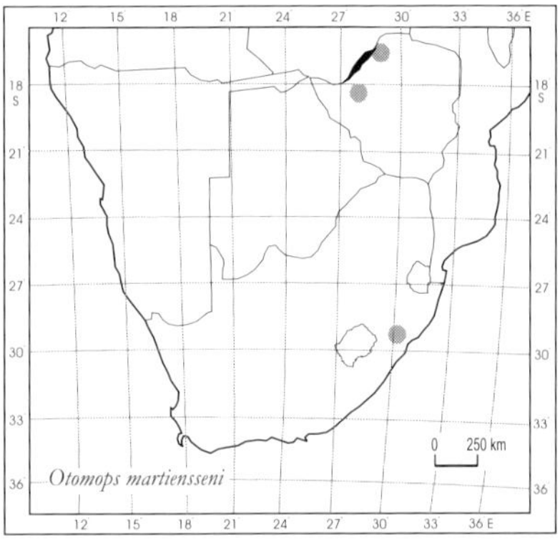

Southern African distribution only

	Males			Females		
	Mean	Range	*n*	Mean	Range	*n*
FA	66	65–68	6	64	62–66	13
Mass	30	26–34	5	29	27–33	13

Taylor (1998b; unpublished)

Guano bats *Tadarida* (Rafinesque 1814)

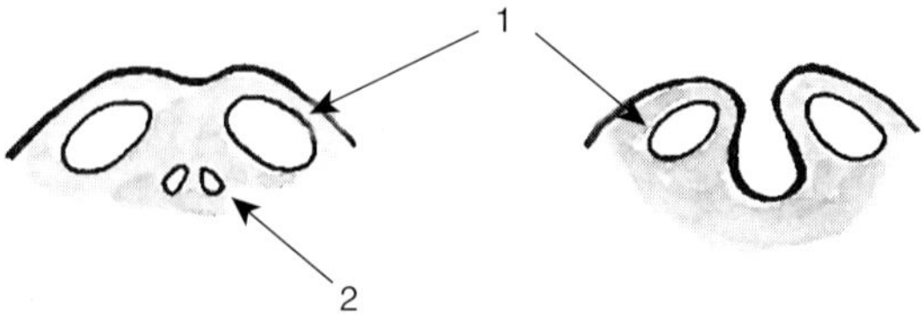

Figure 36 Variation in palatal emargination in free-tailed bats
1. Canine
2. Incisive foramena

The most reliable character for distinguishing members of this genus, the well developed anterior palatal emargination, which is at least as wide as the diameter of the upper incisor (Figure 36), is not very helpful in the field. However, a combination of other characters make these bats distinctive, including their square-shaped ears which are not joined by a band of skin over the nose, the more prominent tragus which is less hidden by the antitragus than in *Mops* and *Chaerephon*, the absence of the post-aural tuft of hairs found in male *Chaerephon*, and the absence of the paler, sparse-haired band across the shoulders found in *Mops*.

Of the four southern Africa species of the genus *Tadarida*, the Egyptian free-tailed bat (*T. aegyptiaca*) is widespread throughout southern Africa; the Madagascar free-tailed bat (*T. fulminans*) occurs marginally in South Africa, as well as throughout Zimbabwe; the giant African free-tailed bat (*T. ventralis*) is known from a few specimens from the former Transvaal (the exact locality of the specimen is unknown), Zimbabwe, and from an isolated record from Mozambique; while the big-eared free-tailed bat (*T. lobata*) is known from a few specimens from Zimbabwe. In spite of the extreme rarity of the two last-mentioned species, recent collecting by Cotterill (1996) suggests that they may be less rare than previously thought. Both species have been taken from urban and suburban parts of the city of Harare, Zimbabwe.

Egyptian free-tailed bat *Tadarida aegyptiaca* (E. Geoffroy 1818)

Dark sooty brown above, just a touch paler below; wing and inter-femoral membrane translucent light brown. The species can be distinguished from other species of this genus by its smaller size (FA <55 mm, as opposed to >55 mm in other species) and its wrinkled lips (the larger Tadarida *species have lips that are not wrinkled).*

HABITAT

They are found in all vegetation associations except forest, favouring open country.

HABITS

Gregarious, but, unlike more colonial species such as the Angolan free-tailed bat and little free-tailed bat, colonies seldom number more than a few dozen. Colonies have a very pungent odour.

FEEDING

Beetles and moths have been recorded in the diet. They are high aerial feeders.

REPRODUCTION

After a gestation of four months (recorded in the Eastern Cape), a single young is born, usually in November or December. Females give birth once a year. Maternity colonies are established by females during summer.

DAY ROOST

Crevices in rocks and trees, behind the bark of dead trees, and commonly in suitable crevices in the roofs of buildings.

Figure 37 Egyptian free-tailed bat roosting

CALL

Both relatively steep-FM and shallow-FM calls have been recorded; the former may apply to more cluttered situations. In open situations, calls have a long duration (12 ms) and narrow bandwidth (20–27 kHz), with a low dominant frequency of 22–24 kHz. Up to three harmonics may be present.

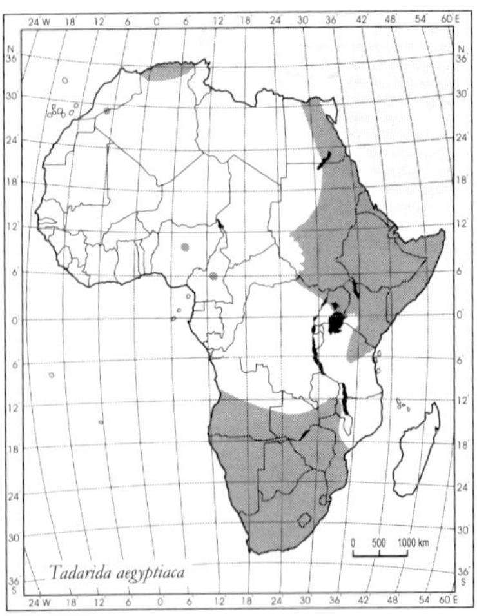

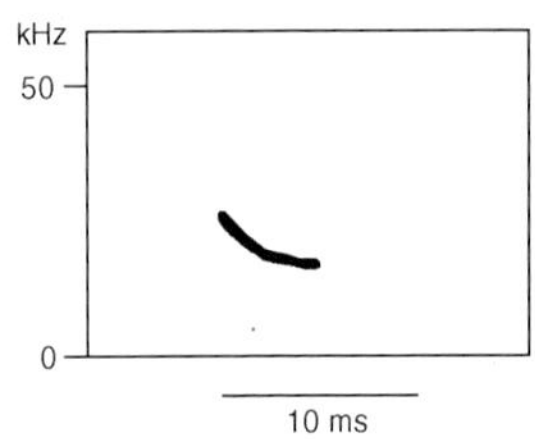

DISTRIBUTION AND STATUS

They occur widely throughout the subregion.

	Males			Females		
	Mean	Range	*n*	Mean	Range	*n*
FA	47	45–54	18	48	46–52	32
Mass	15	11–20	20	17	14–22	32

Sources: Rautenbach (1982); Smithers and Wilson (1979); Taylor (unpublished)

Madagascar large free-tailed bat *Tadarida fulminans* (Thomas 1903)

Larger in size than the Egyptian free-tailed bat, but smaller than the giant free-tailed bat (see Chapter Five). Chocolate brown to reddish above. Males have a fulvous-pinkish ventral colour, while that of younger males and females is pure white. Overall, adult territorial males are 'washed' with fulvous overall (Cotterill, pers. comm.). Lips plain, and not wrinkled like the Egyptian free-tailed bat. The dominant (17 kHz) and minimum (14 kHz) frequencies of their echolocation call are low and just audible.

HABITAT

Associated with crevices within precipices in dolomite, basalt and granite hills in savanna woodland in Zimbabwe.

HABITS

Gregarious, in colonies of usually less than 20.

FEEDING

Moths and beetles. They are high aerial feeders.

REPRODUCTION

Like many other free-tailed bats, females give birth to a single young but are capable of having more than one birth period during the year. However, unlike other species, females give birth not only during the warm wet season, but also during the cool dry season (at least in Zimbabwe).

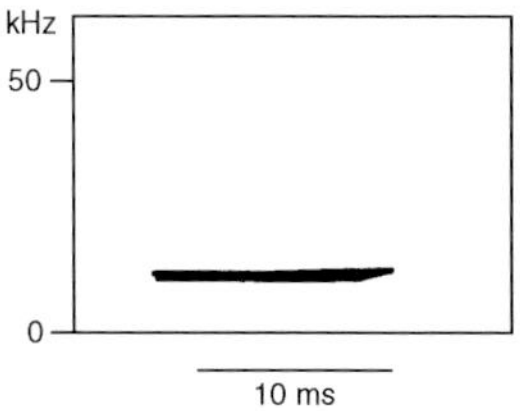

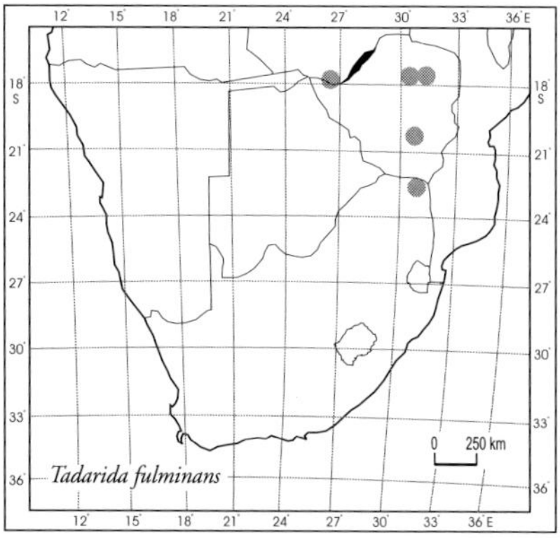

Southern African distribution only

DAY ROOST

Crevices, including cracks in granite, dolomite, or basalt cliffs. An individual was taken from the 18th floor of a high rise building in Harare, and may have been roosting in the roof of the building. The Egyptian free-tailed bat has been frequently collected from or in the vicinity of high rise buildings in the central business district of Durban. It is not inconceivable that bats which are adapted to roosting in crevices in cliffs would find similar roosting situations in the roofs of high rise buildings.

CALL

Long-duration (20 ms), narrow bandwidth (14–27 kHz), multi-harmonic, shallow-FM calls with low dominant frequency of 17 kHz.

DISTRIBUTION AND STATUS

Zimbabwe and the Kruger National Park in South Africa. Recent collecting by Cotterill (In press (b)) shows them to be locally common in suitable habitat in Zimbabwe. Listed as Lower Risk (Near Threatened) in the *1996 IUCN Red List of Threatened Animals*.

	Irrespective of sex		
FA	Mean 58	Range 57–59	*n* 6

Source: Harrison Zoological Institute Museum (HZM) collection

Big-eared free-tailed bat *Tadarida lobata* (Thomas 1891)

This rare, large-sized bat is best recognised by the enormous ears which join together just behind the nostrils, with the tips projecting beyond the snout when laid forward. Unlike the large-eared free-tailed bat, they do not possess a whitish collar, but have a diagnostic white patch of fur between the shoulders.

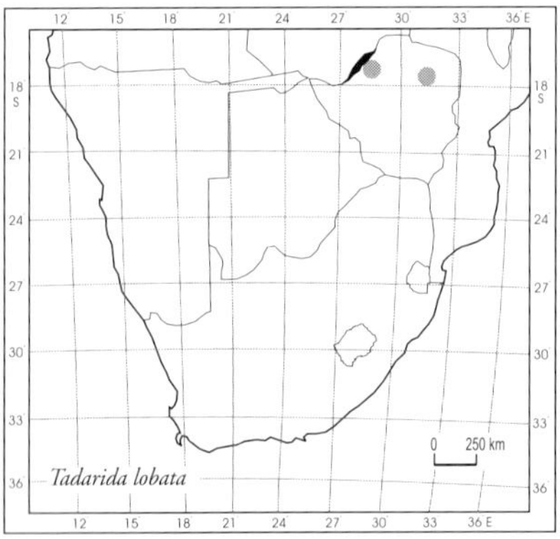

Southern African distribution only

HABITAT, HABITS, FEEDING, REPRODUCTION, AND DAY ROOST

Nothing is known of the biology of this species; probably very similar to *T. fulminans* in terms of high flying habits and inaccessible day roosts.

DISTRIBUTION AND STATUS

Apart from just four known localities in Kenya, in southern Africa it is known from just a few localities in Zimbabwe, including the vicinity of Harare and the shores of Lake Kariba.

	Males		
	Mean	Range	*n*
FA	60	58–62	4

Source: Cotterill (1996b)

Giant African free-tailed bat *Tadarida ventralis* (Heuglin 1861)

This rare, large-sized bat possesses a circular tuft of hair surrounding a bare patch of skin in the middle of the throat (a bare throat patch is sometimes also present in T. fulminans*). Colour above is brown, slightly paler below with a white mid-ventral belly stripe of some 6–7 mm width, reaching from chin to just short of the anus. They can be easily distinguished from the similar-sized* T. lobata *by the absence of a white patch of fur between the shoulders and by the smaller-sized ears (see Chapter Five). They are less easily distinguished externally from* T. fulminans*, but have a larger forearm (>61 mm, as opposed to <60 mm) and skull (see Chapter Five), with a broader inter-dental palate and pronounced and deep basisphenoid pits in the skull.*

HABITAT, HABITS, FEEDING, REPRODUCTION, AND DAY ROOST

Virtually nothing is known of the biology of this species. One specimen was collected from a high rise building in Harare, Zimbabwe, and it is likely to be very similar in habits and day roost to *T. fulminans*. In Kenya, a specimen was collected emerging from the roof of a building.

DISTRIBUTION AND STATUS

Scattered distribution across eastern tropical Africa. In southern Africa, it is known from four localities in Zimbabwe, one from western Mozambique and one from the former Transvaal (exact location of this record unknown). In view of its very restricted distribution, and rarity in museum collections, the giant African free-tailed bat is listed as Lower Risk (Near Threatened) in the *1996 IUCN Red List of Threatened Animals*.

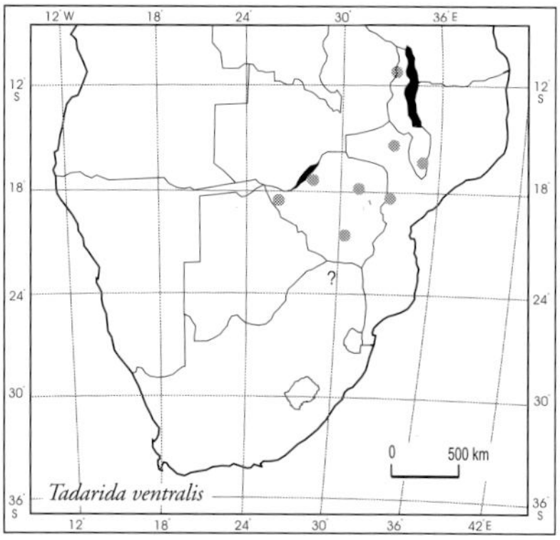

South-central African distribution only

	Males			Females		
	Mean	Range	*n*	Mean	Range	*n*
FA	63	61–64	6	62	61–64	2

Source (for Zimbabwe and East Africa combined): Cotterill (1996b)

FAMILY VESPERTILIONIDAE (PLAIN-FACED BATS)

This is by far the largest and most diverse family of microbats. Their tails are long and fully enclosed in the tail membrane, while their faces are simple and unadorned by noseleafs or slits.

Three subfamilies comprising 10 genera and 29 species are found throughout southern Africa. Members of this family have a widespread global distribution, including cold temperate climates to which they have successfully adapted through their ability to hibernate. In contrast to the tropically-adapted free-tailed bats, of which only half of the 14 southern African species extend their distributions into South Africa, all 10 genera and all but three species of southern African plain-faced bats occur in South Africa, demonstrating a greater tolerance for lower temperatures.

SUBFAMILY MINIOPTERINAE

Long-fingered bats *Miniopterus* (Bonaparte 1837)

In all members of this genus, the second phalanx of the third digit is especially elongated, more than three times longer than the first (see Figure 24), giving the wing its narrow, pointed swallow-like shape when extended in flight. This wing shape is an adaptation for fast flight for feeding in open spaces and for migration. When folded backwards, the elongated third digit protrudes noticeably, giving the wings a 'bent' appearance, hence the Australian name (bent-wing bats) given to this genus. Members of this genus have characteristically small ears and a high, raised forehead (Figure 38). Long-fingered bats roost in densely-packed clusters of individuals in small to large colonies in caves and tunnels, clinging with hind and wing claws, either to the substrate or to other bats (hence, an alternative name for Schreibers's long-fingered bat is the Natal clinging bat) (Figure 39).

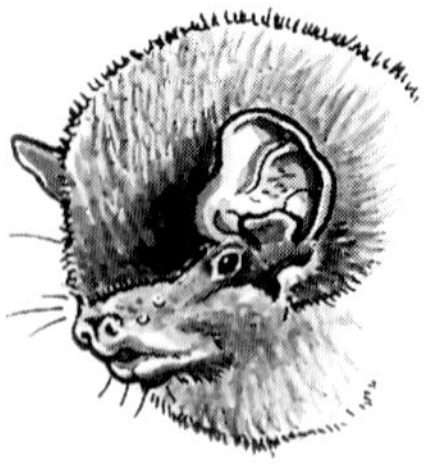

Figure 38 Head of Schreibers's long-fingered bat

Figure 39 Schreibers's long-fingered bat roosting

Long-fingered bats employ a unique method of reproductive delay known as 'delayed embryonic implantation' (see Chapter Three).

In the genus *Miniopterus*, two species are commonly found in southern Africa, the lesser long-fingered bat (*M. fraterculus*), and Schreibers's long-fingered bat (*M. schreibersii*), and a third, the greater long-fingered bat (*M. inflatus*), is represented in southern Africa from just a few localities in Zimbabwe and Namibia.

Lesser long-fingered bat *Miniopterus fraterculus* (Thomas and Schwann 1906)

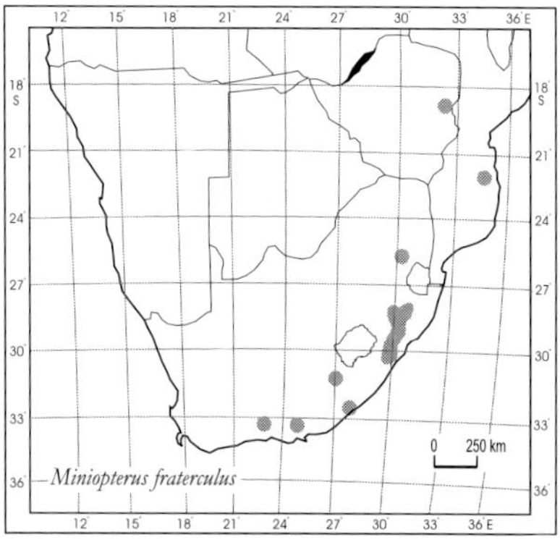

Southern African distribution only

Smaller in size than both the greater long-fingered bat and Schreibers's long-fingered bat (see Chapter Five). Colour in Schreibers's long-fingered bat is just too variable to be useful in distinguishing the two species. Previous keys suggested that the lesser long-fingered bat may be brownish-black or russet, rather than slaty-black in Schreibers's; but at Shongweni Dam in KwaZulu-Natal, the exact reverse seems to apply. Furthermore, brownish and reddish individuals of Schreibers's are often encountered, especially in the moult phase. Plotting a bivariate graph of forearm length versus mass has proved to be a reliable means of distinguishing the two species in the field, since they may often co-occur in the same roost.

HABITAT

From observations in KwaZulu-Natal they occupy a wide range of vegetation types, from drier savanna bushveld to moist mistbelt forests, wherever suitable caves, old mines, or tunnels are present.

HABITS

Colonial, often roosting alongside Schreibers's long-fingered bat, but in much smaller numbers. Probably also make seasonal migrations as described for Schreibers's long-fingered bat.

REPRODUCTION

Like Schreibers's long-fingered bat, this species exhibits delayed embryonic development. Mating occurs in May, two months later than in Schreibers's long-fingered bat. Young are born in November and December, resulting in a gestation of six months.

DAY ROOST

Hollow-roosting: caves, larger rock fissures, overhangs, disused mines and tunnels.

NIGHT ROOST

They probably use night roosts, as does Schreibers's long-fingered bat, but this has not yet been observed.

DISTRIBUTION AND STATUS

Easterly distribution in southern Africa, occurring as far west as the vicinity of George on the coast of the Western Cape. Generally less common with a more restricted distribution than Schreibers's long-fingered bat. Listed as Lower Risk (Near Threatened) in the *1996 IUCN Red List of Threatened Animals.*

	Males			Females		
	Mean	Range	*n*	Mean	Range	*n*
FA	42	42–43	3	42	41–43	2
Mass	–	6.4–10.8	–	–	6.0–9.1	–

Sources: masses – Laycock 1976; FA – Rautenbach (1982); Taylor (unpublished)

Greater long-fingered bat *Miniopterus inflatus* (Thomas 1903) (Plate 7)

In spite of some overlap, it can be distinguished on size (see Chapter Five). Colour is deep chocolate or sooty brown above, a little lighter below.

HABITAT, HABITS, AND DAY ROOST

Presumably similar to Schreibers's long-fingered bat, with which it co-occurs in caves in eastern Zimbabwe.

DISTRIBUTION AND STATUS

Known in southern Africa from just a few localities in eastern Zimbabwe and Namibia.

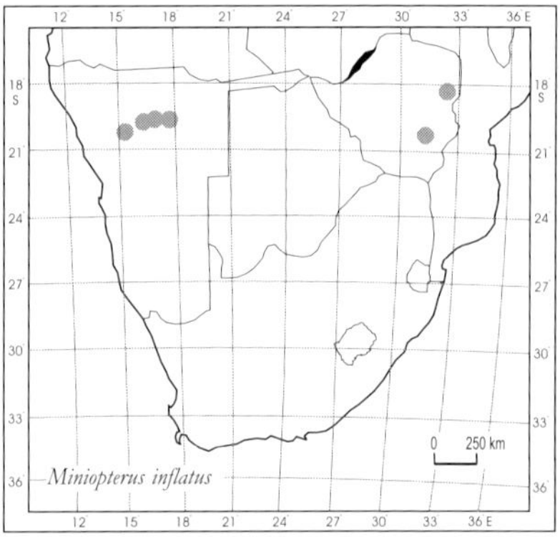

Southern African distribution only

	Irrespective of sex		
	Mean	Range	*n*
FA	47	45–48	4
Mass	14.7	12–17	4

Source: Smithers and Wilson (1979)

Schreibers's long-fingered bat *Miniopterus schreibersii* (Kuhl 1819)

Colour very variable from slaty-grey to brownish-black and russet; the russet colour has been observed during November in the old fur of individuals which were in the process of moulting. As described in the previous account, size is the best means of distinguishing the species, though the echolocation call of the other two species may prove to be distinct from that described below.

HABITAT

They occupy a wide range of vegetation types. Mostly restricted to lower-lying bushveld and coastal regions, but occurring infrequently in the highveld of Gauteng and the Drakensberg range.

HABITS

Cave-dwelling maternity colonies may reach up to 100 000 pregnant females (at De Hoop Guano Cave near Bredasdorp). At least in certain parts of South Africa, they are known to use distinct winter hibernating caves (e.g. highveld of Gauteng and the interior of the Western Cape) and summer maternity caves (e.g. lowveld of the Northern Province, and the De Hoop Guano Cave on the southern Cape coast of South Africa), and to migrate distances of up to 260 km between them.

FEEDING

At Sengwa in Zimbabwe, their diet was found to consist primarily of beetles and moths, while in a separate study at De Hoop Guano Cave in the Western Cape, moths, flies and bugs were predominant in faecal remains.

REPRODUCTION

The species shows delayed implantation of the fertilised blastocyst, resulting in an effective gestation of eight months. The length of the delay period varies with latitude. In KwaZulu-Natal, copulation and fertilisation occurs in March. Implantation of the fertilised blastocyst is delayed during the four months of winter hibernation. Thereafter, in spring, implantation and normal embryonic growth continue for another four months, with young being born in December.

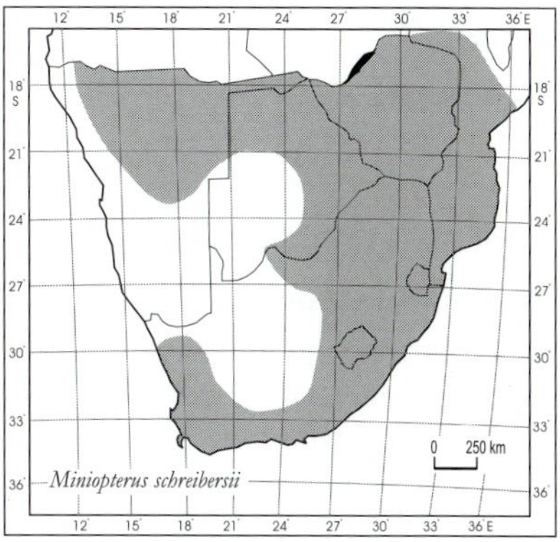

Southern African distribution only

DAY ROOST

Hollow-roosting: caves, disused water and railway tunnels and mine shafts, and substantial cavities in rocks, bridges, dam walls and trees. Maternity colonies require a large cavity with a warm, moist microclimate; at De Hoop Guano Cave, females and their young roost at a temperature of 31° C and a humidity of 91%.

NIGHT ROOST

They are known to use temporary feeding shelters during the night, sometimes making use of open outbuildings, garages and verandahs with overhanging shelter.

CALL

This species has a very broad-band (28–80 kHz), steep-FM call of short duration (3 ms). At a distance of more than a few metres, the higher frequencies (>50 kHz) attenuate rapidly and may not be detected with a bat detector. The dominant frequency of 38–40 kHz coincides with the bottom 'hook' of the characteristically-shaped sonogram where the steeper portion of the call flattens out at the lower end.

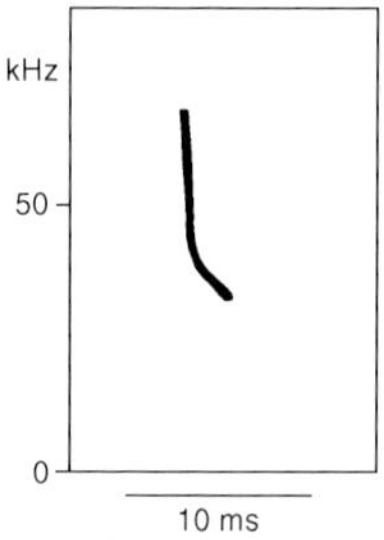

DISTRIBUTION AND STATUS

Widespread and common in the moister southern, eastern and northern regions, but its survival depends on the availability of underground sites having a suitable warm moist microclimate for maternity colonies. The species is classified as Lower Risk (Near Threatened) in the *1996 IUCN Red List of Threatened Animals.*

	Males			Females		
	Mean	Range	*n*	Mean	Range	*n*
FA	45	42–48	62	46	42–52	79
Mass	11	6–13	70	12	8–16	96

Sources: Rautenbach (1982); Taylor (1998b; unpublished)

SUBFAMILY VESPERTILIONINAE

Greater serotine bats *Eptesicus* (Rafinesque 1820)

Recent taxonomic treatments include only one southern African species in this genus, the long-tailed greater serotine bat (*E. hottentotus*). This genus is distinguished from *Pipistrellus* by its larger size, having a mass of over 10 g, and a forearm length exceeding 45 mm, hence the name 'greater serotine bats' is used here to describe this genus, as distinct from the smaller serotine bats which are included in the genus *Pipistrellus.*

Long-tailed greater serotine bat *Eptesicus hottentotus* (A. Smith 1833)

Larger than pipistrelles, FA 46–48 mm; fawn to dark brown above, paler and greyer below, wing membranes always dark brown; ear tragus knife-shaped with rounded point (Figure 27). They have been confused with Schreibers's long-fingered bat, but the absence of an elongated second phalanx of the third finger easily distinguishes them.

HABITAT

Poorly known; frequently collected in mountainous or hilly areas, but also in savanna woodland and in semi-arid conditions in the Western Cape and Namibia.

HABITS

Roost in small colonies of no more than three or four individuals.

DAY ROOST

Caves, rocky crevices and mine tunnels.

CALL

Based on a single room-flown individual, this species has a short duration (2–4 ms), broad-band (20–54 kHz) call with a dominant frequency of 28–32 kHz. A strong harmonic component was usually present, with a second harmonic sometimes present.

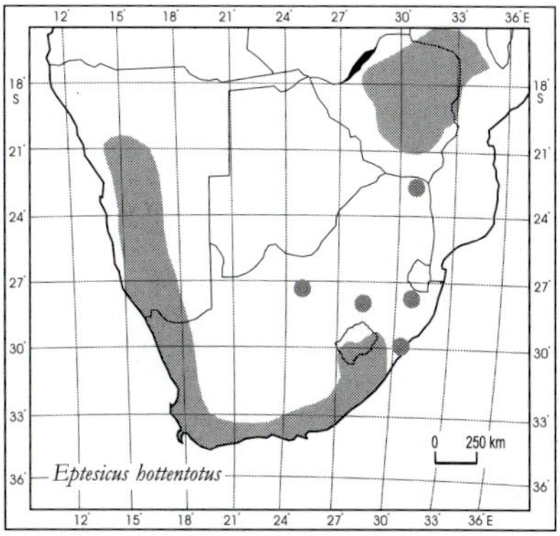

Southern African distribution only

DISTRIBUTION AND STATUS

Records are sparsely distributed over a wide area, although largely absent from much of the interior of the subcontinent. They are apparently locally common in *miombo* woodland on granite hills in southwestern and northeastern Zimbabwe (Cotterill, pers. comm.). Information on their reproduction, general biology, and conservation status is lacking.

	Irrespective of sex		
	Mean	Range	*n*
FA	47	46–48	7
Mass	17	11–20	6

Source: Smithers and Wilson (1979)

Chalinolobus (Peters 1866)

Butterfly bat *Chalinolobus variegatus* (Tomes 1861)

Unmistakable by its pattern of dark reticulations on pale yellow wing and tail membranes; upper body yellowish, cream or pale fawn, paler on head, neck and shoulders, underside off-white or greyish; larger in size than pipistrelles, and very similar in size to Schreibers's long-fingered bat.

HABITAT

Open savanna woodland.

HABITS

Groups of over 10 have been found roosting together. They apparently emit an audible squeak while flying.

FEEDING

Their diet includes moths. They are high flying (18 m) aerial feeders, but are also capable of foraging in 'intermediate clutter' habitats.

DAY ROOST

Poorly known; they have been found under the thatch of an abandoned hut, and concealed between bunches of leaves. Six bats were observed roosting under the canopy of a *Trichelia emetica* tree at Mana Pools in Zimbabwe.

CALL

On take-off and in confined spaces, they produce short (1–2 ms) broad-band, steep-FM calls (30–70 kHz); their search phase calls are longer (5 ms), narrower band, with a more shallow FM shape (33–53 kHz).

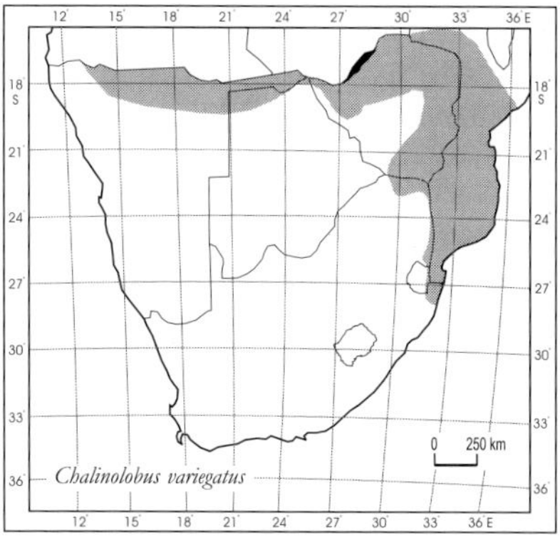

Southern African distribution only

DISTRIBUTION AND STATUS

Widespread in Africa, but rare in collections; in southern Africa, occurring in savanna habitats in the north and northeast, to as far south as the vicinity of Lake St Lucia in KwaZulu-Natal.

	Males			Females		
	Mean	Range	*n*	Mean	Range	*n*
FA	45	–	1	44	44–44	2
Mass	12	11–14	9	13	12–14	3

Sources: Rautenbach (1982); Smithers and Wilson (1979); Taylor (1998b)

Long-eared bats *Laephotis* (Thomas 1901)

Three southern Africa species occur, all of which are rare: the Botswana long-eared bat (*L. botswanae*), De Winton's long-eared bat (*L. wintoni*), and the Namib long-eared bat (*L. namibensis*); the last-mentioned species has been recorded only from a restricted area in Namibia, and is listed as Endangered in the *1996 IUCN Red List of Threatened Animals.* Both the Botswana long-eared bat and De Winton's long-eared bat have been listed as Lower Risk (Near Threatened) in the *1996 IUCN Red List of Threatened Animals.* There has been some confusion over the identity of long-eared bats collected from the Western Cape, Free State, and KwaZulu-Natal, but recent evidence suggests that all these specimens should be classified as De Winton's long-eared bat (Kearney and Taylor 1997). Because of the difficulty in identifying these species, and the poor samples of specimens available to assess intraspecific variation, the Key in Chapter Five should be used with caution, and experts should be consulted concerning identification to species (see Important contacts).

Botswana long-eared bat *Laephotis botswanae* (Setzer 1971) (Plate 8)

Ears are noticeably large, with an average length of 21 mm, though not as long as in the Namib long-eared bat (25 mm); colour buffy brown above, paler below with white-tipped hairs, especially posteriorly; light brown wings. The paler buffy ventral colour and smaller size (FA 33–37 mm) distinguishes it from De Winton's long-eared bat (FA 36–41 mm).

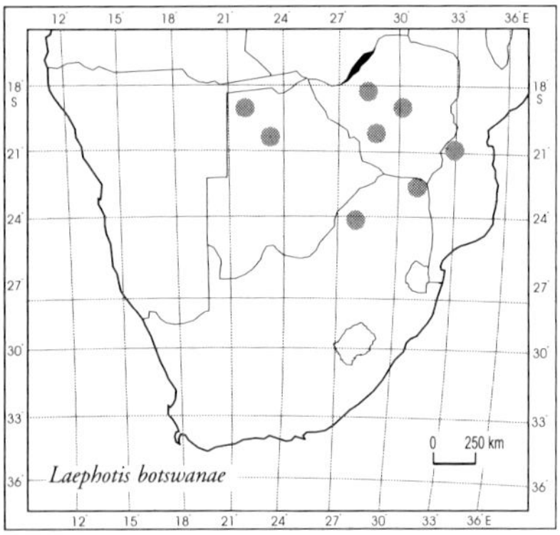

Southern African distribution only

HABITAT

Savanna woodland, near rivers.

HABITS

Unknown. From their ear length, it has been suggested that they may glean their prey from leaves and branches, in the same way as slit-faced bats.

FEEDING

There have been no reported observations of feeding by gleaning, but they have been observed to be aerial feeders, with a diet which includes predominantly moths, with lacewings and beetles occurring in much lower numbers.

CALL

Short duration (5 ms), broad-band (32–55 kHz), steep-FM call having a dominant frequency of 33 kHz. A harmonic may be present.

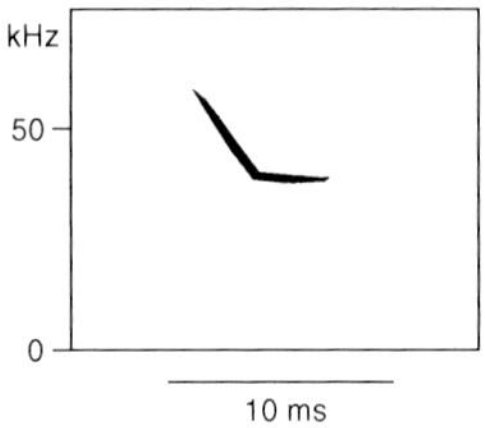

DISTRIBUTION AND STATUS

Scattered distribution in savanna regions of southern Africa. A rare species, listed as Lower Risk (Near Threatened) in the *1996 IUCN Red List of Threatened Animals*. Its taxonomic relationship with the other species of *Laephotis* mentioned above remains to be clarified.

	Irrespective of sex		
	Mean	Range	*n*
FA	36	33–37	7
Mass	5.8	4.5–6.8	7

Source: Smithers and Wilson (1979)

Namib long-eared bat *Laephotis namibensis* (Setzer 1971)

Ears noticeably elongated, 25 mm in length. Distinctly pale buffy brown colouration above, wing membranes olive brown.

HABITAT, HABITS, AND DAY ROOST

Two specimens were collected at a waterhole in the Kuiseb River of the Namib Desert where they were presumably foraging for insects at night. Their day roosts are assumed to be under the loose bark of suitable *Acacia* trees along the river bed.

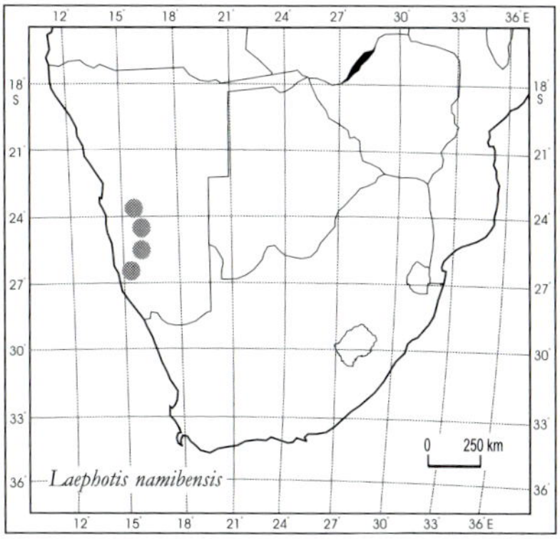

DISTRIBUTION AND STATUS

The entire species is know from just a few specimens from a restricted area of Namibia. They are listed as Endangered in the *1996 IUCN Red List of Threatened Animals.*

	Males			Females		
	Mean	Range	*n*	Mean	Range	*n*
FA	–	–	–	38	38–39	2

Source: Setzer (1971)

De Winton's long-eared bat *Laephotis wintoni* (Thomas 1901)

Colour a shade of mid-brown above, slightly paler below. This species has smaller ears and is not as pale coloured as the Namib long-eared bat, and can be distinguished from the Botswana long-eared bat by its larger size (FA >37 mm) and ventral colour which is less pale buffy than in the Botswana long-eared bat (see Chapter Five).

CALL

In a room-flown individual, very short duration (3 ms), broad-band (17–70+ kHz), steep-FM call having a dominant frequency of 23 kHz. Frequencies exceeding 70 kHz are sometimes present but become rapidly attenuated by the earth's atmosphere.

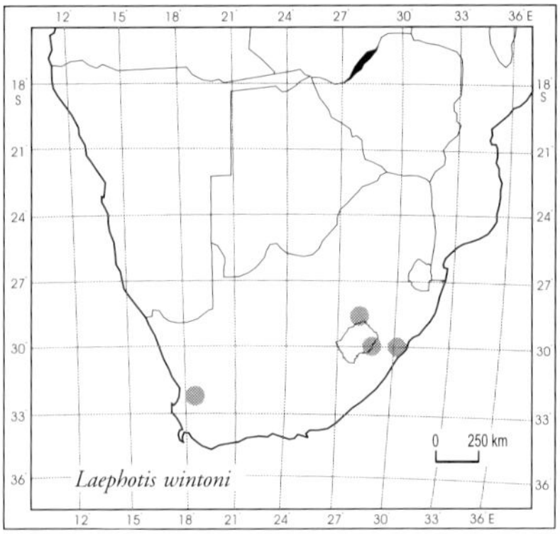

Southern African distribution only

DISTRIBUTION AND STATUS

Specimens from the Western Cape, Free State, Lesotho, and the Richmond area of KwaZulu-Natal have been provisionally assigned to this species. Outside southern Africa they occur only in Ethiopia and Kenya, representing a huge gap in distribution. They are listed as Lower Risk (Near Threatened) in the *1996 IUCN Red List of Threatened Animals.*

	Irrespective of sex		
	Mean	Range	*n*
FA	39	36–41	16
Mass	7	–	1

Sources (total distribution): Taylor (1998b); Watson (1990a); Rautenbach and Nel (1978); Hill (1974a)

Mimetillis (Thomas 1904)

Moloney's flat-headed bat *Mimetillus moloneyi* (Thomas 1891)

This species was only recently reported (Cotterill, In press (a)) to occur in southern Africa, from the Zinave National Park in Mozambique. This genus and its sole species is readily recognisable on external field characters alone. The very short, dark fur, the narrow wing, broad, flat head, short, club-like tragus, and the sharp reduction in the length of the third digit of the wing, combine to separate this species sharply from any other southern African bat. However, since specimens of Scotoecus *strongly resemble this genus in general size and colour and in having a broad muzzle and short tragus, care should be taken to check the reduction in the third digit (which gives the wing a very narrow appearance), which is less than head and body length in* Mimetillus, *but greater than head and body length in* Scotoecus.

	Males			Females		
FA	Mean 28	Range 28–29	*n* 2	Mean 29	Range 28–30	*n* 3

Source (Central and East Africa): MNHN and HZM collections

Hairy bats *Myotis* (Kaup 1829)
Members of this genus are generally identified by their coppery-reddish colour and the 'fluffy' nature of their fur. They are unique in having six pairs of cheekteeth in both the upper and lower jaws, and a pronounced muzzle. Differences between species are based on size, colour and the presence or absence of wing glands (see Chapter Five). Five species occur in southern Africa, Welwitsch's hairy bat (*M. welwitschii*), Temminck's hairy bat (*M. tricolor*), the Angola wing-gland bat (*M. seabrai*), Lesueur's wing-gland bat (*M. lesueuri*), and the rufous hairy bat (*M. bocagei*). The Angola and Lesueur's wing-gland bats are endemic to southern Africa.

Rufous hairy bat *Myotis bocagei* (Peters 1870)
Distinguished on its size (see Chapter Five). Striking, coppery-red colouration above, and underparts paler, off-white washed with red. The ears are black, short and sickle-shaped; wings and tail membrane dark.

HABITAT
Forest and riverine forest habitats within the Kruger National Park.

HABITS
Solitary species.

DAY ROOST
Dense vegetation, in the leaves of palms, terminal leaves of banana plants, or in bunches of bananas.

DISTRIBUTION AND STATUS
Recorded from eastern, northern and central Zimbabwe (records from central and northern Zimbabwe are very recent – Cotterill, pers. comm.), and from Skukuza and Pafuri within the Kruger National Park.

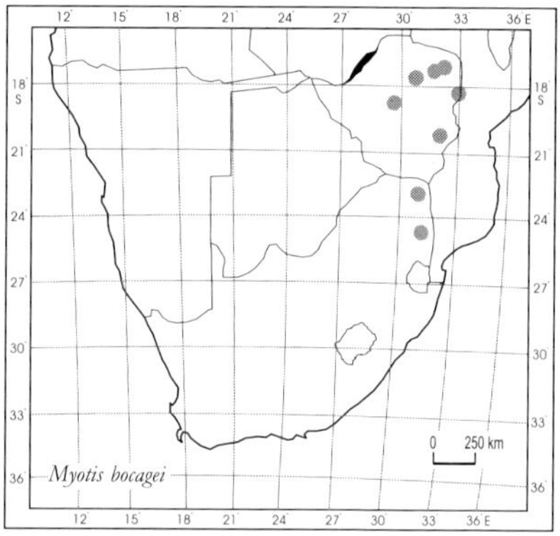

Southern African distribution only

	Males			Females		
	Mean	Range	*n*	Mean	Range	*n*
FA	39	34–41	3	41	39–42	7
Mass	8	6–9	3	9	7–10	7

Source: Rautenbach (1982)

Lesueur's wing-gland bat *Myotis lesueuri* (Roberts 1919)

Slightly larger than the Angola hairy bat. Honey-yellow coloured above, much paler yellowish-white below.

HABITAT AND DAY ROOST
They occupy both mountainous and semi-arid habitats. They are known to roost in rock crevices.

DISTRIBUTION AND STATUS
Endemic to South Africa where they are restricted to the Western and Northern Cape, Free State and Lesotho. Listed as Vulnerable in the *1996 IUCN Red List of Threatened Animals.*

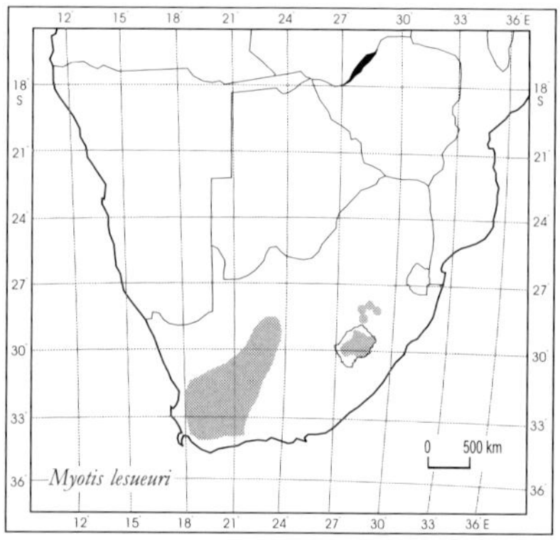

	Irrespective of sex		
	Mean	Range	*n*
FA	38	35–39	49
Mass	7.5	5–10	49

Source: Lynch (1994)

Angola wing-gland bat *Myotis seabrai* (Thomas 1912)

Smallest of the species of Myotis, forearm length being the only reliable means of distinguishing this species from the very similar Lesueur's hairy bat (see Chapter Five). Colour is a dull greyish-yellow, overall darker than Leseur's wing-gland bat. Apparently the presence of wing glands is variable in the species, and may be present on just one wing.

HABITAT

Arid terrain with less than 100 mm annual rainfall.

HABITS

They apparently have a more steady flight pattern than the Cape serotine. At Goodhouse, on the lower Orange River, they were observed gleaning insects from the leaves of orange trees.

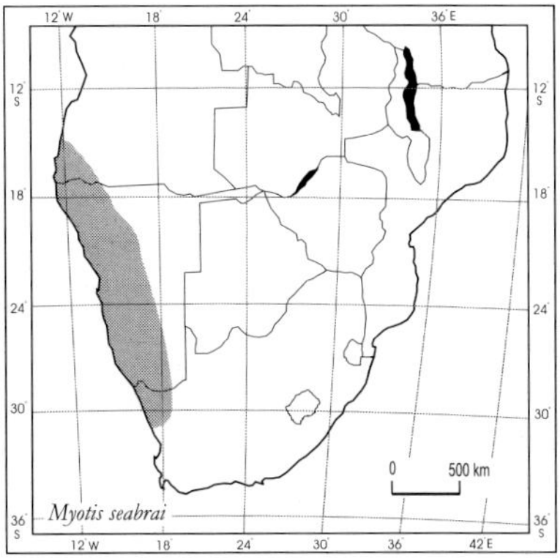

DISTRIBUTION AND STATUS

Endemic to the arid and semi-arid regions of the Northern Cape and Namibia, apart from a narrow extension into southwestern Angola. Listed as Vulnerable in the *1996 IUCN Red List of Threatened Animals.*

	Males		
FA	Mean –	Range 32–33	*n* –

Source: Skinner and Smithers (1990)

Temminck's hairy bat *Myotis tricolor* (Temminck 1832) (Plate 9)

Distinctly larger than both Angola and Lesueur's hairy bats, but smaller than Welwitsch's hairy bat and without the distinctive patterning of the latter; lacking wing glands. Colour is coppery-reddish above, slightly paler below, wing membranes dark brown.

HABITAT

Typically savanna woodland, but extending into more arid areas, as well as mountainous areas in the Drakensberg, wherever suitable roosting caves are present.

HABITS

Gregarious, typically roosting in caves or tunnels in tight clusters comprising a few individuals to thousands in a colony, often together with Schreibers's long-fingered bats. Maximum colony size of 1 400 recorded in KwaZulu-Natal; up to 2 000 bats recorded at De Hoop Guano Cave in the Western Cape during sum-

mer. In parts of South Africa at least, they migrate hundreds of kilometres between summer maternity caves and winter hibernating caves.

REPRODUCTION

Gestation 63 days. A single young is born per year. Mating occurs before hibernation, and sperm are stored in the female reproductive tract during the winter hibernation period, with ovulation and fertilisation occurring in spring.

DAY ROOST

Separate mixed-sex, winter hibernating caves or tunnels and all-female summer maternity caves or tunnels. At De Hoop, they prefer cooler maternity roosts (21° C) than do Schreibers's long-fingered bats (31° C).

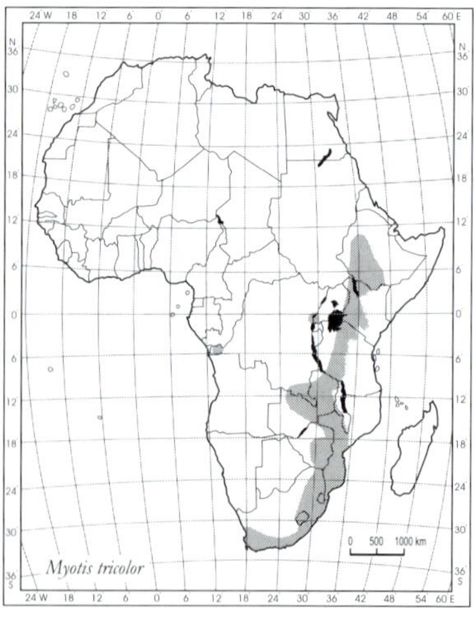

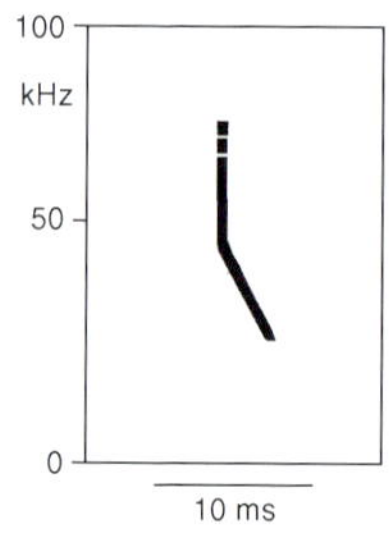

CALL

This species has a very broad-band (33–86 kHz), steep-FM call of short duration (3 ms). While the frequency range is almost identical to that for Schreibers's long-fingered bat, Temminck's hairy bat has a distinctly higher dominant frequency (50 kHz, as opposed to 40 kHz), which coincides with the angle between the steeper and shallower portions of the call on the boomerang-shaped sonogram.

DISTRIBUTION AND STATUS

Fairly common and widespread in eastern parts of the subregion.

	Males			Females		
	Mean	Range	*n*	Mean	Range	*n*
FA	51	–	1	50	41–55	10
Mass	–	–	–	15	12–17	4

Source: Taylor (1998b; unpublished)

Welwitsch's hairy bat *Myotis welwitschii* (Gray 1866)

The largest of the hairy bats; contrasting black spots on orange tail membrane, and contrasting pale orange markings over the bones of the fingers on the dark brown wing membrane, make these rare, solitary bats impossible to confuse with any other bat.

HABITAT

Typically savanna woodland, also grassland-coastal forest mosaic.

HABITS

Unlike the social Temminck's hairy bat, they are solitary.

DAY ROOST

Shrubs and trees; a specimen collected hanging from a low shrub at Craigieburn in southern KwaZulu-Natal was initially mistaken for a dead leaf. A specimen in the Harrison Zoological Museum Institute collection was collected hanging from a cypress tree in a suburb of Harare.

DISTRIBUTION AND STATUS

Scattered distribution in Zimbabwe and northern South Africa, but with outlying southern records from Bloemfontein (Free State) and Craigieburn, near Umkomaas (KwaZulu-Natal).

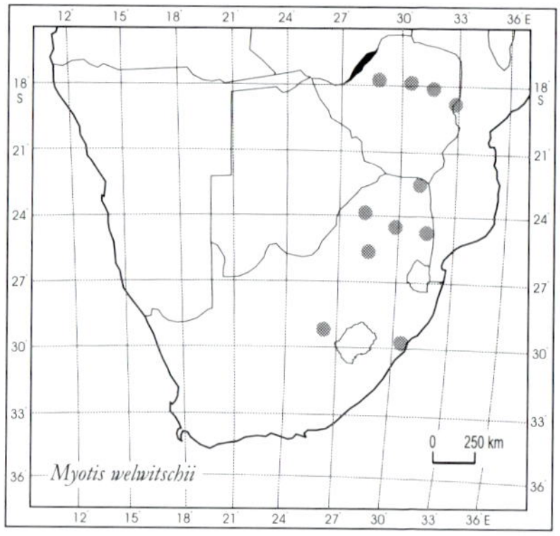

Southern African distribution only

	Irrespective of sex		
	Mean	Range	*n*
FA	55	52–58	6
Mass	14	12–16	6

Source: Smithers and Wilson (1979)

Schlieffen's bat *Nycticeinops schlieffeni* (Peters 1859)

Note: This species was previously known under the genus name, *Nycticeus*.

Similar in appearance and size to pipistrelles, but has a distinctive pale fawn dorsal colour, slightly paler below, contrasting with much darker brown coloured wings and interfemoral membrane, characteristic tragus shape, with a notch on the inner edge and two bumps on the outer edge (Figure 28).

HABITAT

Savanna woodland, often in vicinity of pans or riparian woodland; apparently with a wide habitat tolerance, occurring in areas receiving between 250 mm and 1 500 mm annual rainfall.

HABITS

Apparently solitary in West Africa; no data for southern Africa.

FEEDING

They are aerial feeders, foraging in 'woodland edge' habitats. Diet is dominated by beetles and lacewings, together with, at lower proportions, flies, moths, bugs, and Hymenoptera. In a study at Sengwa in Zimbabwe, the proportion of different insect groups varied throughout the year; lacewings were absent in the hot dry season, but became a dominant food item during the wet season and cold dry season. Diet also varies geographically; Hymenoptera were absent in the diet of Schlieffen's bat at Sengwa but made up 25% of the diet at the Kruger National Park.

REPRODUCTION

Gestation is 11 weeks. Between one and three young are born at a time, with a single birth period per year. Mating occurs before winter, with sperm being stored in both the male and female reproductive tracts. Ovulation and fertilisation occur in spring followed by normal embryonic growth, with young being born around November.

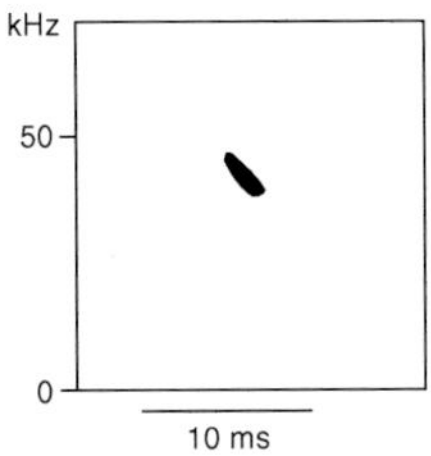

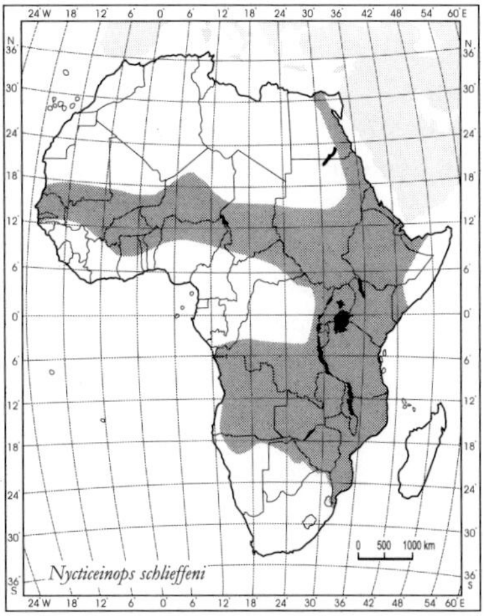

DAY ROOST

In West Africa, they roost in houses, huts, cellars, and crevices in trees; based on label data from specimens from Zambia in the Harrison Zoological Museum collection, a group of three was taken together from a tree.

CALL

Short duration (2 ms), relatively narrow-band (39–47 kHz) steep-FM calls with a dominant frequency of 41 kHz were recorded in Swaziland, while multi-harmonic calls having similar dominant frequency and minimum frequency, but much higher maximum frequency (78 kHz), have been reported in the Kruger National Park and Zimbabwe.

DISTRIBUTION AND STATUS

Widespread and fairly common throughout its indicated range which includes the northern and northeastern parts of the subregion.

	Males			Females		
	Mean	Range	*n*	Mean	Range	*n*
FA	30	28–32	6	32	30–33	29
Mass	5	4.8–5.2	6	5.1	3.5–6.6	35

Source: Rautenbach (1982); Taylor (1998b)

Pipistrelles *Pipistrellus* (Kaup 1829)

Includes Kuhl's pipistrelle (*P. kuhlii*), the rusty bat (*P. rusticus*), the banana bat (*P. nanus*), Rüppell's bat (*P. rueppellii*), Anchieta's bat (*P. anchietai*), Rendall's serotine (*P. rendalli*), the aloe serotine (*P. zuluensis*), and the Cape serotine (*P. capensis*). Owing to uncertainty over its specific status, Melck's serotine is here considered to be a synonym of the common Cape serotine, and is not considered further.

The last-mentioned three species were included in previous treatments within the genus *Eptesicus* (serotine bats), but the current trend, followed here, is to include them within *Pipistrellus*, usually within the subgenus *Neoromicia*, leaving only one southern African species in the genus *Eptesicus*, the larger-sized Hottentot's greater serotine bat (*E. hottentotus*).

Pipistrelles are among the bats most commonly seen at dusk, and can easily be identified to genus because of their relatively small size, compared with the yellow house bats which often forage at the same time, and their laboured, fluttering flight pattern.

Distinguishing between the various species of pipistrelles (the 'little brown jobs', or 'LBJs') is difficult in the field. The most reliable characteristics for distinguishing between pipistrelles – chromosomes and baculum (penis bone) morphology – are certainly not practical for field identification. The species key used relies heavily on features of the skull and dentition; features relating to the incisors, canine, and anterior premolar teeth can sometimes be discerned with care and practice in a live animal, using a magnifying lens.

When more echolocation data become available, bat detectors may prove useful in discriminating between species in the field, e.g. see differences between *P. nanus, P. kuhlii, P. rueppellii, P. capensis* and *P. rusticus* in species accounts to follow. The large degree of bandwidth overlap between species will probably necessitate detailed analysis of sonograms produced from echolocation recordings, or a very well-trained human ear. Until a key using field characters can be developed it is better to refer the 'LBJs' to the experts (see Important contacts). The shape of the ear tragus is also apparently distinctive for each species, but detailed studies of intraspecific variation are required before the reliability of this character can be proven.

A few tips based on fur colour, habitat, and distribution may help guide the reader to identify a particular pipistrelle species. For example, small pipistrelles taken from an unfurling banana leaf are almost certainly banana bats, *P. nanus*. Rüppell's bat (*P. rueppellii*) is recognisable by its pure white undersurface. By far the commonest and most widespread South African 'LBJ' is the Cape serotine (*P. capensis*), a species which is often present as colonies occupying roofs of residential houses throughout southern Africa. It is the only *Pipistrellus* occurring throughout the interior and western regions of South Africa.

Anchieta's bat *Pipistrellus anchietai* (Seabra 1900)

Very similar to Kuhl's pipistrelle. Although distinguishable on the basis of chromosomal number and the shape of the baculum, they are virtually indistinguishable in the field.

HABITAT, HABITS, FEEDING, REPRODUCTION, AND DAY ROOSTS

Assumed to be similar to other species of pipistrelles.

DISTRIBUTION AND STATUS

Scattered distribution in Zimbabwe and northeastern parts of South Africa. Until recently, known in the subregion from a couple of specimens; probably commoner than previously thought as it is likely to have been confused with the commoner Kuhl's pipistrelle in museum collections. Listed as Vulnerable in the *1996 IUCN Red List of Threatened Animals.*

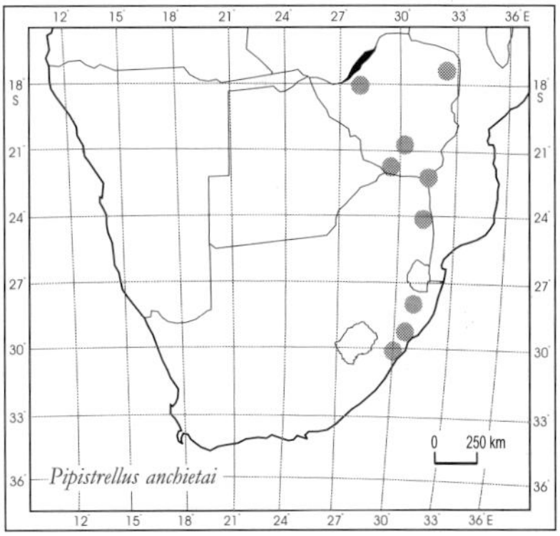

Southern African distribution only

	Males			Females		
	Mean	Range	*n*	Mean	Range	*n*
FA	30	29–30	2	32	32–33	4
Mass	4.3	4.2–4.3	2	4.9	4.4–5.4	3

Source: Taylor (1998b)

Kuhl's pipistrelle *Pipistrellus kuhlii* (Kuhl 1819) (Plate 10)
Dark brown-coloured above, slightly paler below; wing and tail membranes dark, almost black; tragus distinctive, knife-shaped, with a rounded tip (Figure 31); size slightly larger (FA 25–35 mm, mass 4–7 g) than in the rusty bat (FA 27–29 mm, mass 3–4 g). Dominant frequency 50 kHz.

HABITAT
Although it is known from deserts in North Africa, it occupies moister savanna, coastal forest and Afromontane forest habitats in southern Africa.

HABITS
Apparently gregarious, occupying colonies of up to 12.

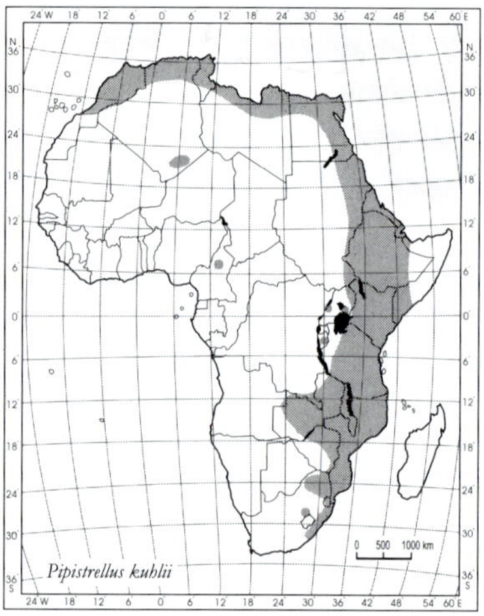

FEEDING AND REPRODUCTION

No data for southern Africa; presumed to be similar to other species of pipistrelles.

DAY ROOST

Poorly known; under bark of trees, in granite crevices, and in roofs.

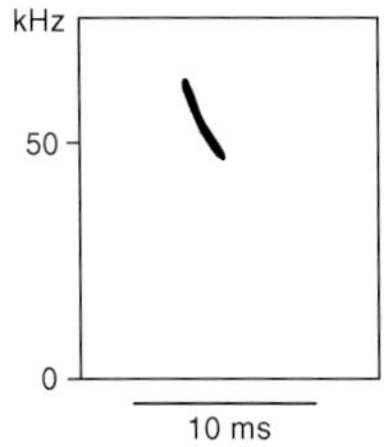

CALL

Short duration (3 ms), fairly broad-band (49–63 kHz) steep-FM call with a dominant frequency of 50 kHz.

DISTRIBUTION AND STATUS

Very wide total distribution, occurring throughout eastern and northern Africa, Europe, the Middle East and India; in southern Africa, occurring mostly in the northeast and east, extending along the coast as far south as East London in the Eastern Cape. Within their range, they are one of the commonest species to be mistnetted at dusk.

	Males			Females		
	Mean	Range	*n*	Mean	Range	*n*
FA	32	31–33	12	33	31–35	8
Mass	5.8	5.1–6.6	12	6.1	5.8–6.6	7

Source: Taylor (1998b)

Banana bat *Pipistrellus nanus* (Peters 1852)

Tiny bat, mass 3–5 g, FA 29–34 mm; ear tragus hatchet-shaped, with abrupt angle on its outer edge (Figure 29); colour variable, from light brown to reddish-brown, somewhat paler than Kuhl's pipistrelle, underparts lighter; wing and tail membranes darker brown compared with body colour.

HABITAT

Typically savanna woodland; they require proximity to water, and banana or *Strelizia* plants in which to roost. Colonies have been located in small banana groves of 10 or more plants in built-up areas of Durban, as well as in the midst of vast expanses of sugar cane monocultures in Swaziland.

HABITS

They roost in groups of up to about six or seven individuals per leaf, but a reasonably small banana grove (30–50 plants) may contain up to 100 individuals. According to research done by D.C.D. and M. Happold in Malawi, males use house roosts during the breeding season.

FEEDING

They are aerial feeders, foraging on the edges of woodland habitats. Diet comprises mostly small beetles and moths, with fewer numbers of flies.

REPRODUCTION

In the above-mentioned Malawi study, mating was promiscuous, as groups of bats comprising different individual males and females occupied unfurling banana leaves at different times during the mating season. After the young were born, and during the period of lactation, males roosted separately from the females. Gestation is 10 weeks, females almost always give birth to twins (occasionally single young), and there is only one pregnancy per year. After mating, sperm is stored in both the male and female reproductive tracts.

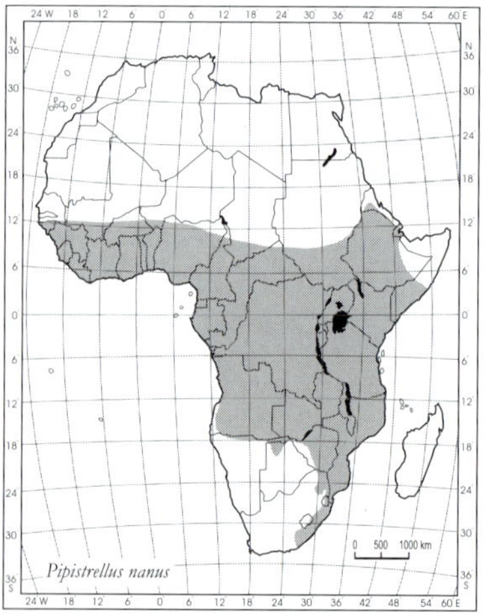

DAY ROOST

Typically the unfurling leaves of banana or *Strelizia*, but they also roost in the thatch of huts, or rafters or other crevices within roofs.

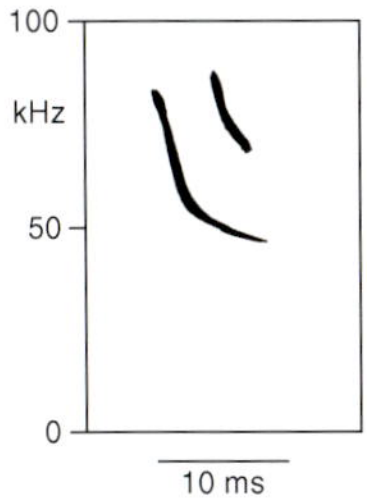

CALL

Short duration (2–3 ms), steep-FM search call with a range of 68–89 kHz, and a dominant frequency of 74 kHz; a further, broader-band call has also been recorded.

DISTRIBUTION AND STATUS

Widespread throughout Africa south of the Sahara; in southern Africa, occurring in the extreme north of Namibia, the Okavango Delta in Botswana, most of Zimbabwe, Mozambique, and the northern and eastern parts of South Africa, extending along the east coast southwards to the Bedford District of the Eastern Cape.

	Males			Females		
	Mean	Range	*n*	Mean	Range	*n*
FA	31	29–38	44	32	30–34	43
Mass	3.8	3–5	23	4.4	3–6	28

Sources: Rautenbach (1982); Smithers and Wilson (1979); Taylor (1998b)

Rüppell's bat *Pipistrellus rueppellii* (Fischer 1829) (Plate 11)

Their pure white underparts, which are clearly demarcated from the brownish-grey to sepia-coloured upper parts, make this species easily distinguishable from other pipistrelles; also slightly larger, with a mass of around 7 g; ear tragus long and knife-shaped with a rounded point.

HABITAT

Riverine conditions, associated with major river systems such as the Nile, Zambezi, Limpopo, and recently collected on the lower Orange River at Augrabies. Often found in open areas devoid of trees (Harrison, pers. comm.).

HABITS

Apparently congregate when feeding.

FEEDING

Diet comprises beetles, moths, lacewings and flies.

DAY ROOST

Apparently known to emerge from cracks in the ground in the Middle East (Harrison, pers. comm.). A specimen collected in Zambia in the Harrison Zoological Institute collection was taken from behind a notice board.

CALL

Short-medium duration (4–8 ms), broad-band (40–70 kHz) steep-FM call, having a dominant frequency of 45 kHz, and up to two harmonics.

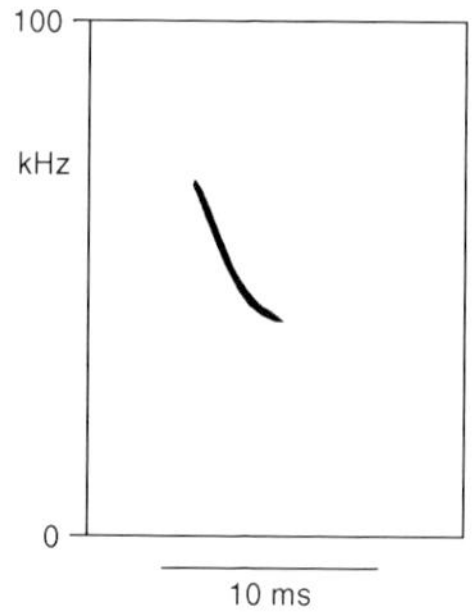

DISTRIBUTION AND STATUS

Uncommon, but widely distributed in Africa and the Middle East; in southern Africa, known from the Okavango Delta in Botswana, Zimbabwe, Mozambique, and the Kruger National Park in South Africa, with an isolated record from Augrabies on the Orange River in the Northern Cape (Churchill, pers. comm.).

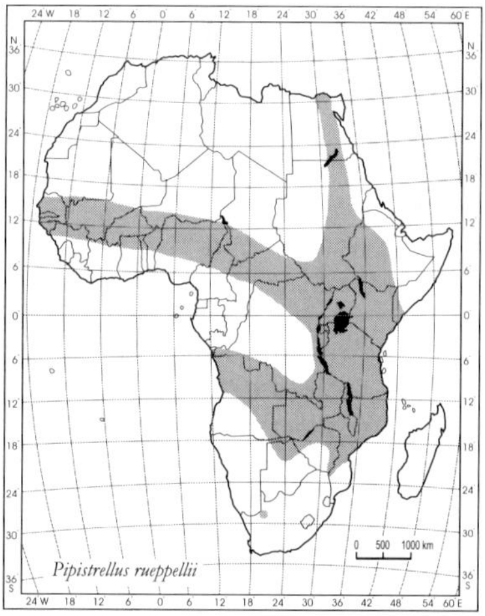

Southern African distribution only

	Irrespective of sex		
	Mean	Range	*n*
FA	34	30–39	9
Mass	7.1	6.8–7.3	5

Sources: masses – Smithers and Wilson (1979); FA – HZM collection (entire range)

Rusty bat *Pipistrellus rusticus* (Tomes 1861)

Colour pale rust above, paler below, with dark, almost black tail and wing membranes; southern Africa's smallest bat species, with a mass of only 3–4 g, and FA 27–30 mm; tragus is sickle-shaped with a rounded point (Figure 30).

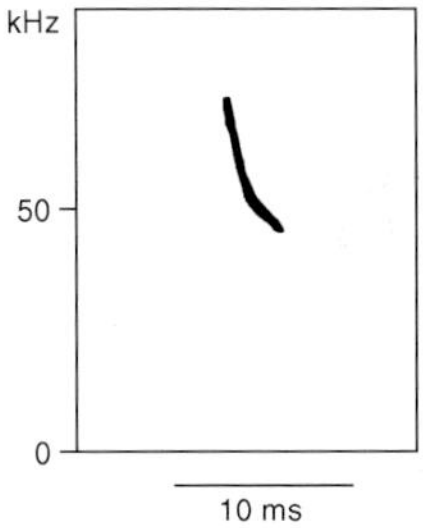

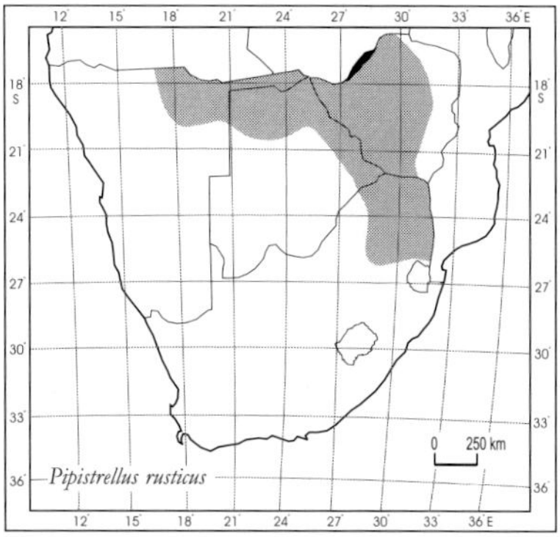

Southern African distribution only

HABITAT

Savanna woodland, often riverine associations.

FEEDING

Aerial feeder, foraging on woodland edges. Diet includes beetles and moths.

REPRODUCTION

Gestation is 12 weeks. Females invariably give birth to twins once a year. Following mating, sperms are stored in both the female and male reproductive tracts.

DAY ROOST

According to label information on a specimen in the Harrison Zoological Institute collection, a group of three bats was taken together from a hollow tree in Zambia.

CALL

Short duration (3 ms), broad-band (46–69 kHz) steep-FM call with a dominant frequency of 51 kHz, i.e. very similar to Kuhl's pipistrelle.

DISTRIBUTION AND STATUS

Limited mostly to the northern parts of the subregion.

	Males			Females		
	Mean	Range	*n*	Mean	Range	*n*
FA	28	27–29	19	29	27–30	23
Mass	4.0	3.1–4.8	10	4.4	3.6–4.9	19

Source: Skinner and Smithers (1990); Rautenbach (1982)

Subgenus *Neoromicia*

The following three species were previously included in the genus *Eptesicus* on the basis of having four upper cheekteeth, i.e. lacking the tiny fifth premolar present in other pipistrelles. This character is difficult though not impossible (with a hand lens) to ascertain in the field (see Chapter Five), and is unreliable across the entire worldwide range of this genus. The distributions and distinguishing external features listed below need to be considered, e.g. the translucent white wings of Rendall's serotine, and the low dominant frequency of the Cape serotine (around 40 kHz, as opposed to 50 kHz or more in the banana bat, Kuhl's pipistrelle, and rusty bat). In most cases, it will take an expert, or a lot of practice, to reach a positive identification. As indicated previously, Melck's serotine is no longer recognised to be a good species, and is not considered further.

Cape serotine *Pipistrellus capensis* (A. Smith 1829)

Very variable in colour; dark to light, on average yellowish-brown coloured above, yellowish-white below, wing and tail membranes blackish-brown; females larger than males; overall larger (FA 29–38 mm, mass 4–10 g) than the aloe serotine. Dominant frequency 38–41 kHz.

HABITAT

Very broad habitat tolerance, from forests to deserts, although not recorded from the coastal Namib Desert. Often found in suburban situations, although rare in the greater Durban area where they may compete for roof space with little free-tail bats which seem better able to tolerate the hot, humid coastal conditions.

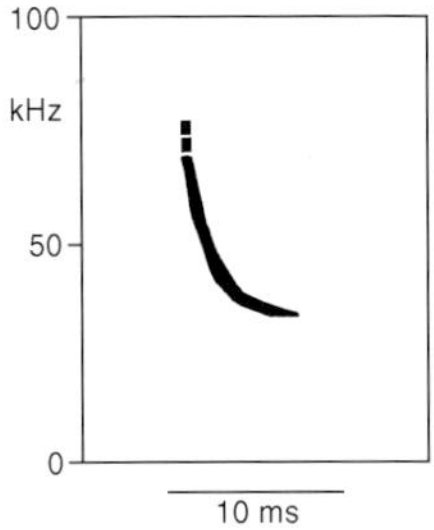

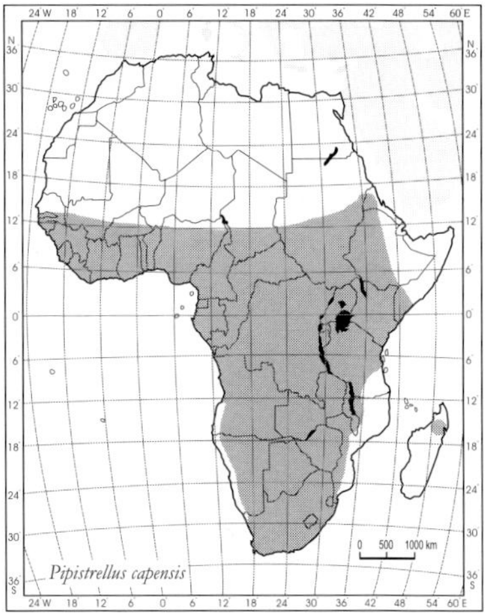

HABITS

Roost tucked together in small groups of two or three. They are attracted to insects around lights, and are one of the most commonly seen small vesper bats flying at dusk, jinking and turning in flight, along set flight paths, at a height of some 10–15 m.

FEEDING

They are aerial, 'woodland edge' feeders. Diet comprises mostly beetles and lacewings, with fewer moths, bugs and flies. In a study at Sengwa, Zimbabwe, the proportion of different insect groups varied throughout the year; for example, lacewings were absent in the hot dry season, but became a dominant food item during the wet season and cold dry season. Diet also varies geographically; moths were poorly represented in the diet of Cape serotines at Sengwa, but made up 22% of the diet in the Kruger National Park.

REPRODUCTION

Gestation is 12 weeks. Females give birth once a year, around late November, to twins, triplets, or even quadruplets. Sperm storage occurs in both males and females.

DAY ROOST

Tucked into crevices in rocks, under the bark of trees, at the base of aloe leaves, and frequently in roofs.

CALL

Short-medium duration (4–8 ms), very broad-band (33–80 kHz), steep-FM calls, with a dominant frequency of 38–41 kHz, which coincides with the 'hook' at the lower end of the sonogram. The higher frequencies (>65 kHz) attenuate rapidly at distances of more that a few metres.

DISTRIBUTION AND STATUS

A very common, widespread, and hardy species, occupying even inhospitable, semi-arid habitats in the Karoo where very few other species of bats survive.

	Males			Females		
	Mean	Range	*n*	Mean	Range	*n*
FA	32	29–35	43	34	31–38	38
Mass	6.1	4–10	40	7.6	4–10	40

Sources: Rautenbach (1982); Taylor (1998b)

Rendall's serotine *Pipistrellus rendalli* (Thomas 1889)

Their distinctive, translucent white wings distinguish them from other pipistrelles; also slightly larger size (FA 34–37 mm, mass 8–9 g).

HABITAT

Known from well-watered situations, including Mana Pools in Zimbabwe, Okavango Delta, and a pan within the Lala palm-veld in Zululand, South Africa.

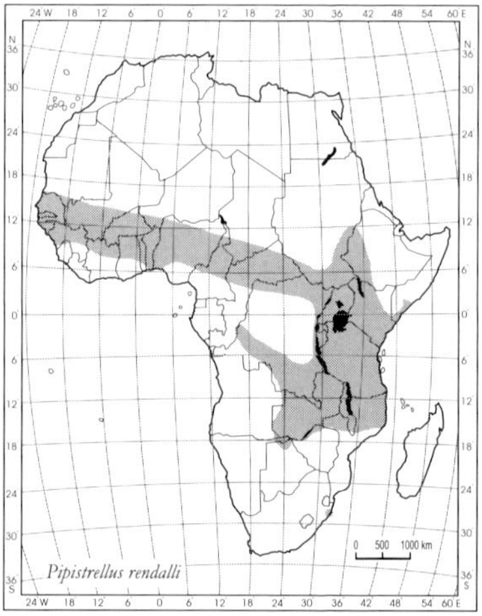

HABITS

They forage close to the ground, usually less than 2 m.

DAY ROOST

Not known for the subregion; in West Africa they roost in roofs of buildings and in dense foliage of trees and bushes.

DISTRIBUTION AND STATUS

Occur marginally in the northern parts of southern Africa, although their recent discovery at Bonamanzi Private Game Reserve in northern KwaZulu-Natal represents an extreme southerly range extension. Listed in the Specially Protected category in the KwaZulu-Natal Conservation Ordinance of 1999.

	Males			Females		
	Mean	Range	*n*	Mean	Range	*n*
FA	36	34–37	3	38	–	1
Mass	8.3	7.9–8.5	3	–	–	–

Source: Taylor (1998b)

Aloe serotine *Pipistrellus zuluensis* (Roberts 1924)

Similar in colour to the Cape serotine, slightly paler above; slightly smaller (FA 30–33 mm), without the posterior occipital 'helmet' of P. capensis *(best observed from a prepared museum specimen, although an experienced person should be able to detect this character in the field by gently feeling the back of the head). This species is limited to savanna woodland regions and is not found in grassland, semi-arid, or montane habitats where the Cape serotine occurs.*

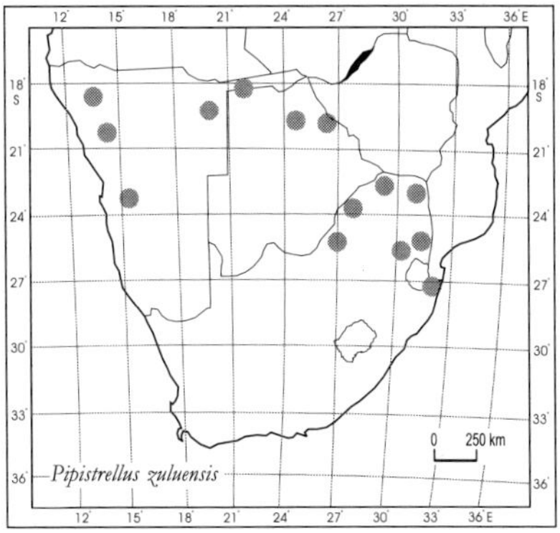

HABITAT
Savanna woodland.

HABITS
Poorly known; they associate with the Cape serotine.

FEEDING
Aerial 'woodland edge' feeders. Diet includes beetles and moths.

DAY ROOST
Presumably under the bark of trees; poorly known.

DISTRIBUTION AND STATUS
Not as common as the Cape serotine, with scattered distribution in Namibia, northern Botswana and northern South Africa.

	Males			Females		
	Mean	Range	*n*	Mean	Range	*n*
FA	31	31–31	3	32	30–33	4
Mass	4.6	3.0–7.0	3	4.3	4.0–4.8	3

Source: Rautenbach (1982)

Lesser house bats *Scotoecus* (Thomas 1901)

Until recently, only one species, Thomas's lesser house bat, was known to occur in southern Africa. The presence of the dark-winged lesser house bat at Zinave National Park in Mozambique was recently noted by Cotterill (In press (f)). The genus can be distinguished from other small-sized plain-faced bats on the basis of a single, prominent, procumbent incisor, an obviously broad rostrum (which is about half of the greatest skull length), and, in males, an elongated penis which measures 25–40% or more of forearm length (7–14 mm). Lesser house bats are separated from house bats on their smaller size, with FA <46 mm (>46 mm in *Scotophilus*), and on the short, blunt tragus (compared with the long, tapering tragus of *Scotophilus*).

Thomas's lesser house bat *Scotoecus albofuscus* (Thomas 1890)

This rare vespertilionid bat is easily identified by its distinctive white wings, broad, flat head, club-shaped tragus, and small size. Colour of the St Lucia specimen (see below) was a light buffy brown (a darker brown in a second juvenile specimen); forearms, tail membrane and membrane close to the body a dark brown to blackish colour, contrasting strongly with the white wings. However, it can be confused with another rare species, Mimetillus moloneyi, *which also has pale wings, a broad flat skull and club-shaped tragus, but from which it can be distinguished by having wings which are not particularly narrow (see species account for* M. moloneyi).

HABITAT

Not known; the second specimen known from southern Africa was collected recently from a holiday resort in the village of St Lucia, close to coastal forest bordering the St Lucia Estuary in KwaZulu-Natal.

DAY ROOST

Unknown in southern Africa; a male and female in the Harrison Zoological Institute Museum collection were taken together in a *Hyphaene* palm in Tanzania, and an association with *Hyphaene* palms was also described for this species in West Africa by Rosevear.

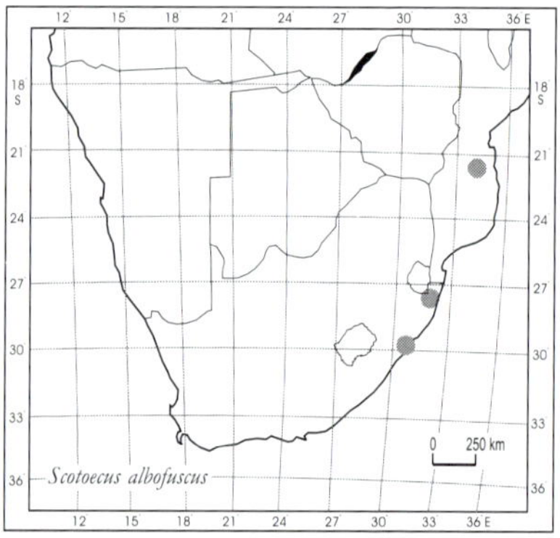

Southern African distribution only

DISTRIBUTION AND STATUS

Although previously known from only a single record in Mozambique, a pregnant female was recently collected in St Lucia, KwaZulu-Natal by F. and N. Radford of the Durban Bat Interest Group. Just before this book went to print, a further specimen, a juvenile male, was collected from a garden in the Durban suburb of Yellowwood Park. Listed as Lower Risk (Near Threatened) in the *1996 IUCN Red List of Threatened Animals.*

	Irrespective of sex		
	Mean	Range	*n*
FA	–	30–31	–

Source: Hill (1974b) (total range)

Dark-winged lesser house bat *Scotoecus hirundo* (De Winton 1899)
This bat is clearly distinguished from the above species by its dark-coloured wings. There is some uncertainty about the number of species of dark-winged lesser house bats. The accuracy of the name, Scotoecus hirundo, *used here, depends on whether it is synonymous or not with other described species (see Taylor and Van der Merwe 1999 for a discussion). The presence of this species at Zinave National Park in Mozambique was recently established by Cotterill (pers. comm.).*

	Irrespective of sex		
	Mean	Range	*n*
FA	–	35–38	–

Source: Hill (1974b) (total range)

House bats *Scotophilus* (Leach 1821)
Members of this genus are distinguished from other vespertilionid bats by their large size, long tragus, and pug-like face (see Chapter Five).

Three species are found in southern Africa: the yellow house bat (*S. dinganii*), the lesser yellow house bat (*S. viridis*), and the rare giant yellow house bat (*S. nigrita*); the last-mentioned species is known from southern Africa from a few specimens in Zimbabwe, Botswana, and Mozambique (although it occurs elsewhere in Africa), and is listed as Lower Risk (Near Threatened) in the *1996 IUCN Red List of Threatened Animals.*

Yellow house bat *Scotophilus dinganii* (A. Smith 1833) (Plate 12)
Distinguished from the lesser yellow house bat by its larger size (FA 50–58 mm, mass 14–42 g), the sleek texture and olive colour of the upper parts, and the canary-yellow colour of the belly. However, caution should be taken when identifying non-adult yellow house bats where the yellow colour may not be fully developed. Non-adult status can be determined in a bat by holding the wing up to the light: a thin pale line across the end of the long finger bones indicates a non-adult in which the cartilaginous epiphysis has not yet calcified. Although colour seems so far to be a consistent character in southern African specimens, this is very variable in West Africa, where S. dinganii *frequently have white bellies, and* S. viridis *sometimes have yellow bellies. Thus, wherever there may be a*

conflict between colour and size, size should be regarded as the more important character for identification.

HABITAT

Savanna woodland and coastal forests, often associated with houses, and preferring lower altitudes (<1 000 m), although apparently present only during summer on the central plateau (1 200–1 500 m) of Zimbabwe.

HABITS

Gregarious, occurring in small clusters of up to 12. Readily enter torpor during winter, and are quite docile in captivity.

FEEDING

Aerial 'intermediate clutter' feeder. Diet comprises mostly medium-sized beetles, as well as a few numbers of bugs, flies, termites, moths, and lacewings.

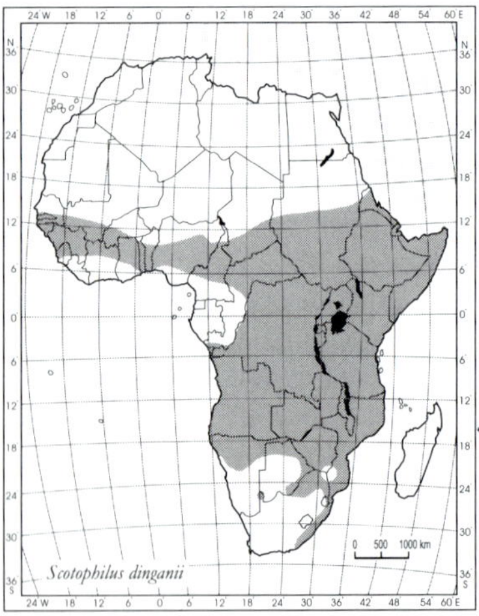

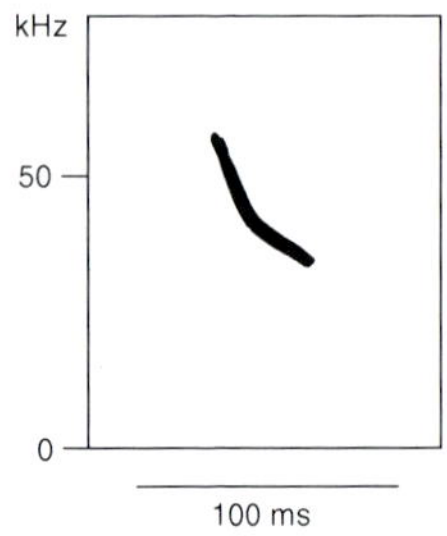

REPRODUCTION

Between one and three young are born at a time, with a single birth per female per year. In KwaZulu-Natal, young are born between November and January.

DAY ROOST

Typically, the roofs of houses where small groups of six or less individuals usually huddle together between the brickwork and rafters.

CALL

Short-medium duration (5–10 ms), fairly broad-band (29–54 kHz) with dominant frequency of around 35 kHz. Call consists of steep-FM followed by shallower-FM components, giving a 'boomerang-shaped' sonogram. A harmonic may be present.

DISTRIBUTION AND STATUS

Common and widespread in the eastern and northern parts of the subregion, extending southwards along the east coast as far south as Port Elizabeth and inland to the vicinity of Craddock. Isolated record from Kalahari Gemsbok National Park.

	Males			Females		
	Mean	Range	*n*	Mean	Range	*n*
FA	55	51–58	52	54	50–58	43
Mass	26	15–36	29	27	14–42	25

Source: Rautenbach (1982); Smithers and Wilson (1979); Taylor (1998b)

Giant yellow house bat *Scotophilus nigrita* (Schreber 1774)

The largest of the yellow house bats, with a forearm exceeding 70 mm. The fur is sleek and short, in colour, dark brown to brown tinged with yellow, red or grey, whitish below in juveniles, yellow below in adults.

DISTRIBUTION AND STATUS

In southern Africa, known from only three localities in Botswana, eastern Zimbabwe and Mozambique. Elsewhere in Africa, known from just a few scattered localities in Central and West Africa and Sudan. Listed as Lower Risk (Near Threatened) in the *1996 IUCN Red List of Threatened Animals*.

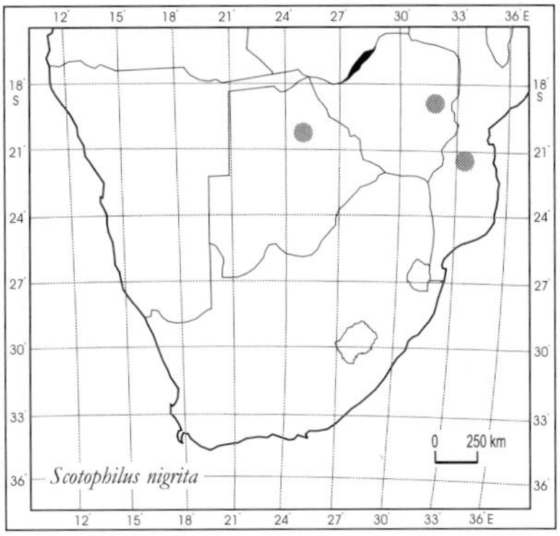

Southern African distribution only

	Males		
	Mean	Range	*n*
FA	–	70–80	–

Source (total range): Rosevear (1965); HZM collection

Lesser yellow house bat *Scotophilus viridis* (Peters 1852)

Note: In the past *viridis* has been included in the species *borbonicus* and *leucogaster*, but is now regarded as separate from both of these.

Smaller than the yellow house, with FA of 44–52 mm, mass 13–35 g (compare with previous); upper colour light to dark yellowish-brown, similar to yellow house bat, but underparts white, or greyish-white, without yellowish wash.

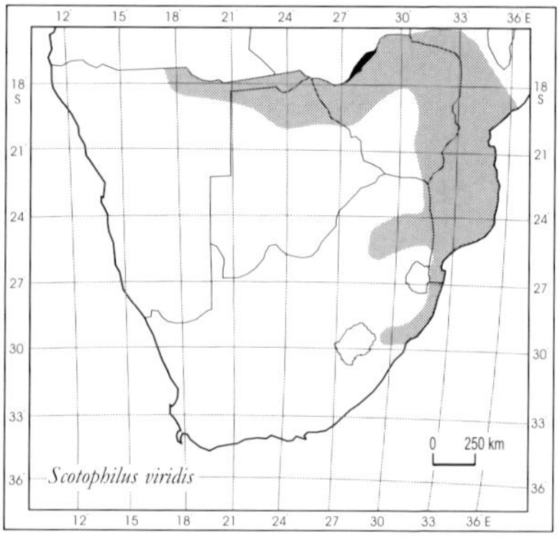

Southern African distribution only

HABITAT

Typically associated with lower-lying savanna woodland or riverine habitats, but recent specimens from KwaZulu-Natal were collected from a wattle stand in open grassland in the Drakensberg (1 800 m) as well as from near the town centre of Pietermaritzburg, suggesting an association with houses.

HABITS

Little known; roost in small colonies.

FEEDING

Aerial, 'intermediate clutter' feeder. In one study, diet comprised beetles, moths, and lacewings, with lower numbers of flies and bugs. A more detailed study revealed a more broad diet with seven orders of insects represented.

REPRODUCTION

Gestation is effectively 7.5 months, due to delayed implantation of the fertilised blastocyst. Unlike in the case of *Miniopterus schreibersii*, where implantation is delayed throughout all of the winter hibernation period, in *S. viridis*, implantation is delayed for half of the winter period, and implantation and normal embryo growth occurs during the latter half. Females give birth to between one and three young in November and December.

DAY ROOST

Tree hollows in bushveld; also probably buildings, although this has not been directly observed.

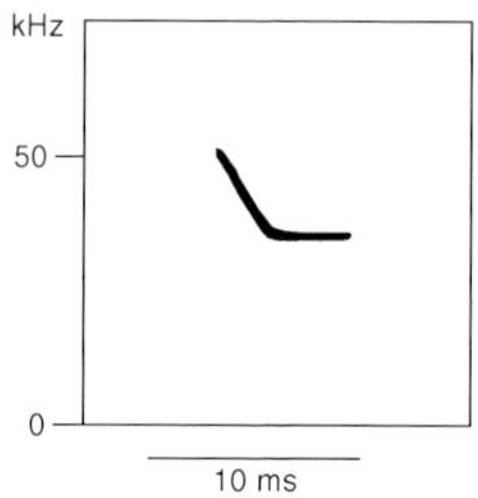

CALL

Calls of varying bandwidth (40–70 kHz, or 34–59 kHz) and duration (5–10 ms) have been reported by studies at the Kruger National Park and Sengwa (Zimbabwe) respectively.

DISTRIBUTION AND STATUS

Common, but slightly less widespread than the yellow house bat, occurring in northern and eastern parts, as far south as Pietermaritzburg, and inland to Garden Castle in the Drakensberg Park.

	Males			Females		
	Mean	Range	*n*	Mean	Range	*n*
FA	49	44–52	13	49	46–51	5
Mass	16	13–18	5	25.8	15–35	4

Source: Rautenbach (1982); Taylor (1998b)

SUBFAMILY KERIVOULINAE

Woolly bats *Kerivoula* (Gray 1842)

These bats are easily recognised by their long, soft, woolly, upstanding hair, the tips of which are curled, giving them a grizzled appearance; the high rounded forehead (more obvious from a prepared skull than on the live animal); the broad, funnel-shaped ears, having a deep emargination below the tip, and the long (9 mm) thin, tapering tragus. Another characteristic feature is the conspicuous fringe of inwardly curving hairs on the outer margin of the tail membrane. Woolly bats frequently roost during the day in abandoned birds' nests (usually those of species of weavers or sunbirds).

Two species are recognised in southern Africa, the Damara woolly bat (*K. argentata*) and the lesser woolly bat (*K. lanosa*), which can be distinguished on external characters.

Damara woolly bat *Kerivoula argentata* (Tomes 1861)

Colour and size as in Key in Chapter Five; ear tragus without the inner basal notch found in the lesser woolly bat.

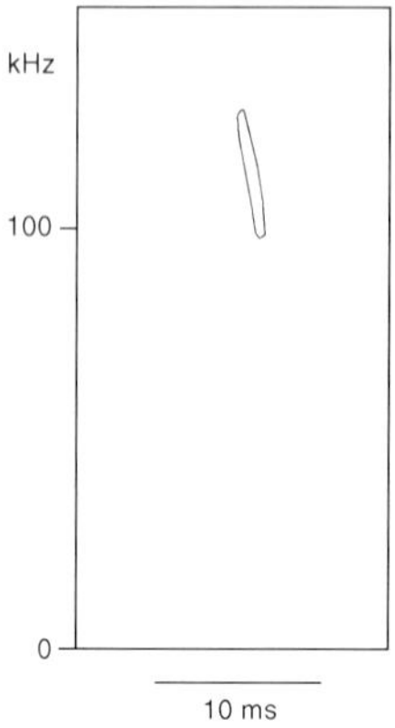

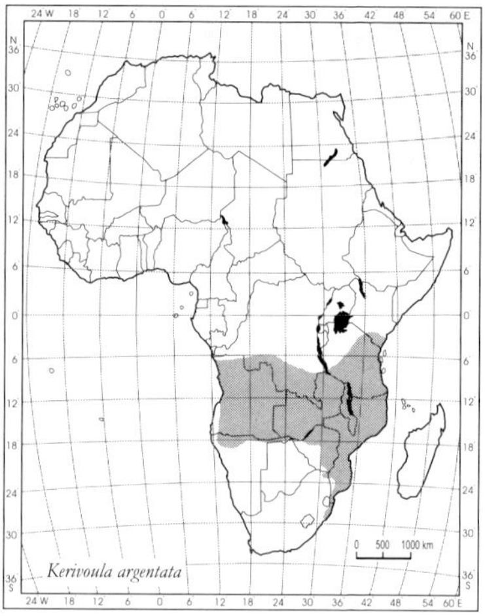

HABITAT

Riverine situations in savanna woodland.

HABITS

Roost singly or in small groups of up to six individuals. Slow, fluttering flight, close to ground.

DAY ROOST

Typically found in the disused nests of masked weavers and spectacled weavers; also outside walls of rondavels, under eaves, rough bark of trees, or in clusters of dead leaves on trees.

CALL

This species has a unique echolocation call among vespertilionid bats, with a low intensity, very high frequency range (85–120 kHz) and dominant frequency of 90–118 kHz, of very short duration (2 ms).

DISTRIBUTION AND STATUS

Wide but scattered distribution in the eastern and extreme northern parts of the subregion; elsewhere, occurring also in East Africa and southern parts of Central Africa. A specimen in the Harrison Zoological Institute Museum collection was washed onto Kings Beach, Port Elizabeth. Since this record falls well outside its known range, it could have been carried along the coast by the sea from farther north (although judging from the good quality of the specimen and the absence of skin-slip, the corpse had not decomposed badly).

Although the individual could conceivably have dispersed by flying, the species is not known to migrate. It is also of course possible that the species does, or did, occur in the area but has not been collected there subsequently. Recently an old but authentic 1926 record of a specimen from close to Durban was uncovered in the American Museum of Natural History, New York, by F.P.D. Cotterill.

	Irrespective of sex		
	Mean	Range	*n*
FA	37	36–41	6
Mass	7.6	6–9	6

Source: Smithers and Wilson (1979)

Lesser woolly bat *Kerivoula lanosa* (A. Smith 1847)

Colour and size as in Key in Chapter Five; ear tragus having inner basal notch not found in the Damara woolly bat.

HABITAT

Similar to above species.

HABITS

Presumably similar to above species. Slow, fluttering flight, close to ground.

DAY ROOST

Also recorded from disused nests of masked weavers, as well as the nest of a scarlet-chested sunbird.

DISTRIBUTION AND STATUS

In southern Africa, scattered records are known from extreme northwestern and southeastern Botswana, eastern Zimbabwe, the Northern Province, northern KwaZulu-Natal, and three localities in the Eastern Cape.

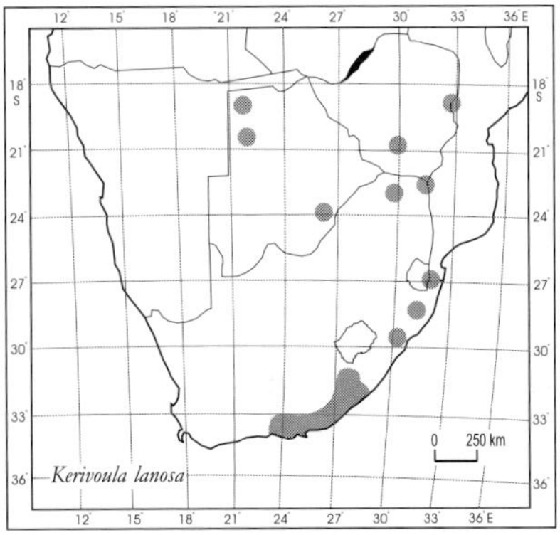

Southern African distribution only

	Irrespective of sex		
	Mean	Range	*n*
FA	32	30–36	8
Mass	6.8	–	4

Source: masses – Smithers (1971); forearms – Roberts (1951)

FAMILY NYCTERIDAE (SLIT-FACED BATS)

Members of this family all possess a vertical facial slit which houses noseleaves, only visible when the slit is open. Echolocation calls are emitted through the nostrils, and the noseleaves play an important part in enhancing echolocation performance, and making these bats extremely manoeuvrable, being able to forage within dense bush. Slit-faced bats are easily recognisable from their relatively large ears, up to 37 mm in the case of the common slit-faced bat. They all have long tails ending at the tail membrane in a characteristic T-shape supported by a bony spur. Spurs termed calcanea support the tail membrane at the ankles.

Slit-faced bats *Nycteris* (G. Cuvier and E. Geoffroy 1795)

Five species occur in southern Africa. The common slit-faced bat (*Nycteris thebaica*) is very widespread in southern Africa, and is by far the most likely to be encountered, while the other four species tend to be sparsely distributed or occur only marginally in southern Africa. The hairy slit-face bat (*N. hispida*) is known from isolated records in northern KwaZulu-Natal, Mozambique and Zimbabwe; the large slit-faced bat (*N. grandis*) occurs throughout northeastern Zimbabwe and adjacent parts of Mozambique; Wood's slit-faced bat (*N. woodi*) occurs in Zimbabwe and in the northern parts of the Kruger National Park, and is listed as Lower Risk (Near Threatened) in the *1996 IUCN Red List of Threatened Animals*; while the greater slit-faced bat (*N. macrotis*) is known only from the Lake Kariba area of Zimbabwe. Two specimens collected in Mozambique in 1965 and previously referred to as Vinson's slit-faced bat (*N. vinsoni*), have recently been reclassified as *N. macrotis*.

Large slit-faced bat *Nycteris grandis* (Peters 1865)

The largest slit-faced bat, as indicated in the Key in Chapter Five; reddish-brown in colour, lighter underneath.

HABITAT

Well developed evergreen, riverine forests.

HABITS

Colonial, in groups of up to six.

FEEDING

They feed by gleaning their prey off branches or the ground, but are also capable of taking their prey in mid-air whilst flying slowly (slow aerial feeding). Their diet includes not only insects (orthopterans, beetles and moths), but larger vertebrate prey such as frogs, fish and birds, as well as bats.

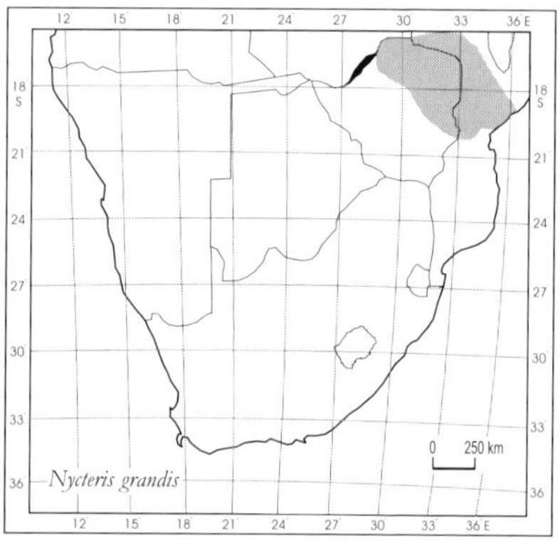

Southern African distribution only

DAY ROOST

Hollow trees, culverts, rocky overhangs, and water towers.

NIGHT ROOST

Open buildings and presumably other sheltered situations.

CALL

More intense than the common slit-faced bat, detectable within 1.5 m of the bat detector; short duration (1–3 ms), multi-harmonic, with up to four peaks and a maximum bandwidth of 17–117 kHz.

DISTRIBUTION AND STATUS

Occurs throughout northeastern Zimbabwe and adjacent parts of Mozambique; a geographically isolated subspecies occurs within the equatorial rain forests of West and Central Africa.

	Males			Females		
	Mean	Range	*n*	Mean	Range	*n*
FA	64	–	1	63	–	1
Mass	39	36–42	2	–	–	–

Source: Skinner and Smithers (1990)

Hairy slit-faced bat *Nycteris hispida* (Schreber 1774) (Plate 13)

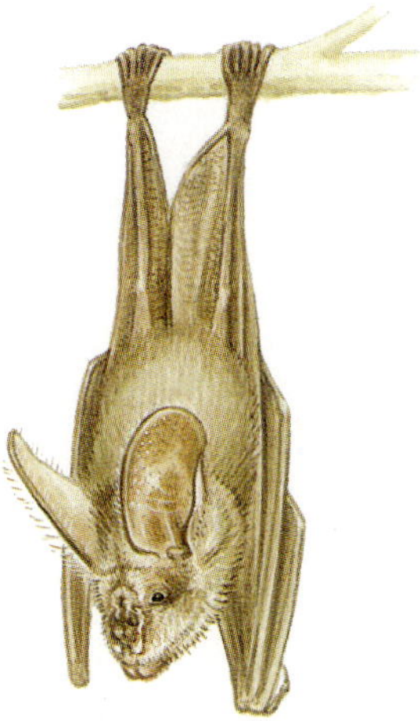

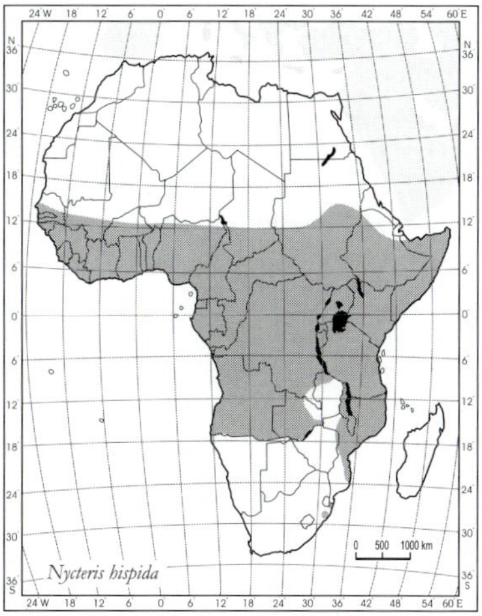

Sepia-brown above, much paler below, ears and wing and tail membranes dark blackish-brown; ears distinctly smaller (18–25 mm) than all other species of slit-faced bat. The tragus is half-moon-shaped and the upper incisors are three-lobed (see Chapter Five).

HABITAT

Wide range of savanna and forest vegetation associations, not occurring in pure grassland or arid habitats.

HABITS

Solitary, or occurring in colonies of up to 20.

DAY ROOST

Low dense bushes, thatched huts, holes in termitaria, hollow trees and caves.

NIGHT ROOST

They have been observed entering houses whilst foraging, presumably using them as feeding stations.

DISTRIBUTION AND STATUS

Widespread and common south of the Sahara; however, in southern Africa, the hairy slit-faced bat is known only from a few isolated records in northern KwaZulu-Natal, Mozambique, and Zimbabwe.

	Males			Females		
	Mean	Range	*n*	Mean	Range	*n*
FA	39	37–43	8	42	41–42	4
Mass	6.5	6–8	3	6	–	1

Sources (West and southern Africa combined): Taylor (1998b); MNHN collection

Greater slit-faced bat *Nycteris macrotis* (Dobson 1876)

Smaller in size than the large slit-faced bat, but larger than other slit-faced bats. It is similar to Wood's slit-faced bat in having a half-moon-shaped tragus and two-lobed upper incisors; however, it can be distinguished from Wood's slit-faced bat by its larger size (FA 45–50 mm, as opposed to 37–42 mm), and brownish (as opposed to greyish) colour.

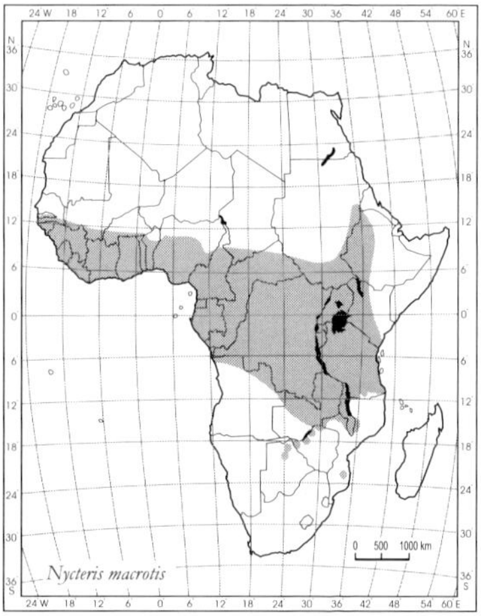

DISTRIBUTION AND STATUS

Although they have a wide but scattered distribution throughout sub-Saharan Africa, the species has been recorded in southern Africa from the Zambezi Valley in Zimbabwe, the south bank of the Save River in Mozambique, and from Serowe and Savuti in Botswana (Cotterill, In press (d)).

	Males			Females		
	Mean	Range	*n*	Mean	Range	*n*
FA	49	47–50	8	50	46–53	11
Mass	13	10–17	7	13	10–16	7

Source (Senegal): MNHN collection

Common slit-faced bat *Nycteris thebaica* (E. Geoffroy 1813) (Plate 14)
By far the commonest encountered species of this genus. The conspicuously long ears (28–37 mm) *make this species readily identifiable (see Chapter Five), together with the finger-like ('pyriform') shape of the tragus (Figure 12); colour buffy brown above, buffy or off-white below; rufous colour phase individuals are quite often encountered; hair long and soft.*

HABITAT
Typically open savanna woodland, as well as dense coastal forest, often associated with buildings.

HABITS
Colonial: colony size varies from a few individuals up to 600 at Drotsky Caves in Botswana. They are very manoeuvrable in flight, being able to forage close to the ground and in dense vegetation.

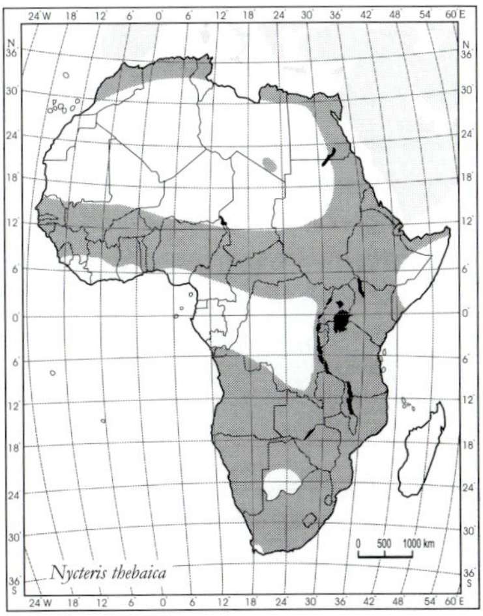

FEEDING
Egyptian slit-faced bats are predominantly gleaners but are also capable of slow aerial feeding. Their long ears are acutely tuned to acoustic signals generated by their prey, for example scuffling sounds of non-flying arthropods moving on the ground, and the courtship calls of male crickets and other noise-producing insects. Based on a study at Mkuzi Game Reserve, their diet com-

prises predominantly orthopteran insects (e.g. katydids, crickets, and grasshoppers), and spiders which are gleaned from branches or from the ground, as well as beetles, cockroaches, moths, and lacewings. Dietary studies reveal large geographical and seasonal flexibility in the diet of this species, e.g. spiders, flies and termites may be either absent in the diet or a dominant food item, depending on their availability at any time.

REPRODUCTION

After a gestation of five months, a single young is born. Females give birth once a year in November. There is no known reproductive delay mechanism.

DAY ROOST

Hollow roosting: caves, mine adits, road culverts, tree hollows, aardvark holes.

NIGHT ROOST

Thatched hides and rondavels, open buildings.

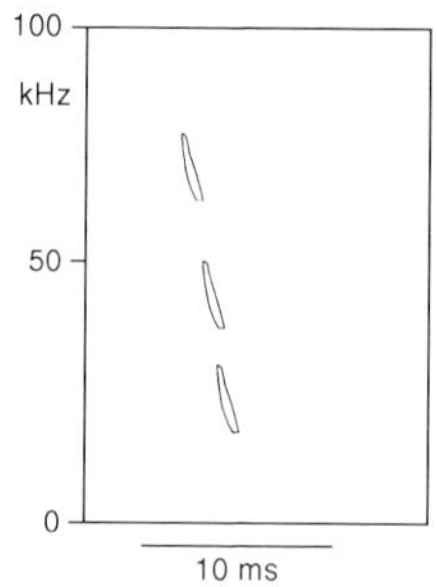

CALL

These bats have characteristically low intensity, short duration (2 ms), multi-harmonic, broad-band steep-FM calls (approximately 20–90 kHz, with peaks at 22 kHz, 44 kHz and 66 kHz), which can only be picked up within 0.5 m of the bat detector.

DISTRIBUTION AND STATUS

Very widespread in southern Africa, except for semi-arid parts of Botswana, and open highveld grasslands of Gauteng, Mpumalanga, the Free State, and Lesotho.

	Males			Females		
	Mean	Range	*n*	Mean	Range	*n*
FA	46	37–51	41	47	42–51	42
Mass	10.5	9–14	52	11.5	7–15	45

Sources: Rautenbach (1982); Smithers and Wilson (1979); Taylor (1998b)

Wood's slit-faced bat *Nycteris woodi* (K. Andersen 1914)

A small species, similar in size to the hairy slit-faced bat but with much larger ears (28–34 mm, as opposed to 18–25 mm); see under account for greater slit-faced bat for differences between these two species.

HABITAT

Savanna woodland.

DAY ROOST

Rocky fissures, buildings.

NIGHT ROOST

No information available but they presumably use night roosts as do other members of this genus.

DISTRIBUTION AND STATUS

In southern Africa, known only from Zimbabwe and the northern parts of the Kruger National Park. Recent collecting (Cotterill 1996a) has shown them to be more widespread in the south and north of Zimbabwe than previously indicated. Outside southern Africa, they are known only from Zambia. Listed as Lower Risk (Near Threatened) in the *1996 IUCN Red List of Threatened Animals.*

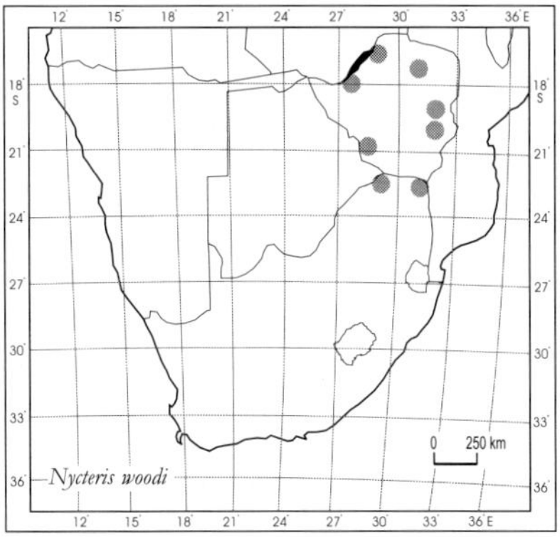

Southern African distribution only

	Males			Females		
FA	Mean 38	Range 37–39	*n* 7	Mean 39	Range 38–40	*n* 5

Source: Smithers and Wilson (1979)

FAMILY RHINOLOPHIDAE (HORSESHOE BATS)

This family of small to medium-sized bats is found worldwide, except for the Americas and polar regions. Only one genus is present. They have adapted to temperate climates through their ability to hibernate, and, in some if not all species, to delay reproduction through the process of sperm storage. It is not surprising then that all 10 southern African species extend their distributions into South Africa, as opposed to only half of the species from the more tropically-adapted Family Molossidae (free-tailed bats).

Horseshoe bats *Rhinolophus* (Lacépède 1799)

Ten species occur in southern Africa: Hildebrandt's horseshoe bat (*R. hildebrandtii*), Rüppell's horseshoe bat (*R. fumigatus*), Geoffroy's horseshoe bat (*R. clivosus*), Darling's horseshoe bat (*R. darlingi*), Cape horseshoe bat (*R. capensis*), bushveld horseshoe bat (*R. simulator*), Dent's horseshoe bat (*R. denti*), Swinny's horseshoe bat (*R. swinnyi*), Lander's horseshoe bat (*R. landeri*), and peak-saddle horseshoe bat (*R. blasii*).

The Key in Chapter Five is based mostly on external characters which can be measured on a live specimen, principally size, and the shape of various components of the horseshoe, such as the connecting process, lancet and sella. Another important character, the presence of a minute upper anterior premolar in the toothrow behind the canine, should be discernible in the field on a live bat with careful examination using a hand lens. Echolocation frequencies are also included in the key as the constant frequency (CF) component can easily be determined from a tuneable (i.e. heterodyne) bat detector in the field, either in a handheld bat or bats flying in a cave.

Peak-saddle horseshoe bat *Rhinolophus blasii* (Peters 1867)

Similar to Lander's bat in size and having a pointed connecting process, but distinguished from this species by its long pale-coloured fur, triangular-shaped lancet (as opposed to concave and narrowly pointed), absence of under-arm (axillary) tuft of stiff hairs in males, and the unshortened first phalanx of the fourth finger (see Chapter Five). Their echolocation call is unknown.

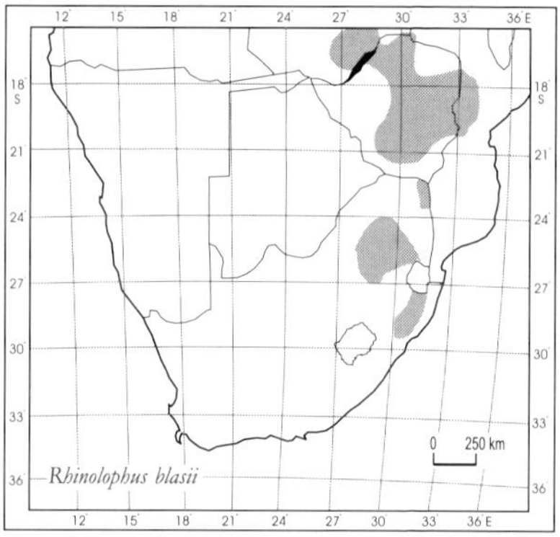

Southern African distribution only

HABITAT, HABITS, AND DAY ROOST

Found in savanna habitats where they roost in caves in small groups of two to three individuals.

DISTRIBUTION AND STATUS

They have a widespread but scattered distribution, occurring as far north and east as southern Europe and southwestern Asia, and as far south as KwaZulu-Natal. In southern Africa, they occur in Zimbabwe, the Northern Province, Mpumalanga, Gauteng, and KwaZulu-Natal. Lower Risk (Near Threatened) in the *1996 IUCN Red List of Threatened Animals.*

	Males			Females		
	Mean	Range	*n*	Mean	Range	*n*
FA	45	41–47	17	46	43–47	16
Mass	4.2	2.1–7.5	3	2.6	2.1–3.0	2

Source: Rautenbach (1982)

Cape horseshoe bat *Rhinolophus capensis* (Lichtenstein 1823)

Colour dark brown above, paler fawn-grey below; connecting process blunt; it is distinguished from the similar-sized Geoffroy's and Darling's horseshoe bats (which also have blunt connecting processes) by having the anterior premolar in the toothrow, separating the canine and posterior premolar, visible with a hand lens.

HABITAT

Occupies a range of habitats associated with coastal caves around the coast of the Eastern Cape and Western Cape provinces.

HABITS

Occurs in cave-dwelling colonies of up to thousands. They often roost together with Geoffroy's horseshoe bat, but in separate clusters. Small-scale migrations of 10 km have been noted.

FEEDING

They feed on the wing whilst flying slowly in dense vegetation (slow aerial 'clutter forager'), and also feed by 'perch-hunting' (hanging on a convenient perch and ambushing passing prey).They may be capable of gleaning although this has not been directly observed. Their diet consists mainly of beetles.

REPRODUCTION

Gestation lasts three to four months. Females give birth once a year to a single offspring. Viable sperm are stored in the male's own reproductive tract throughout winter hibernation when normal sperm production ceases, and not in the female tract as occurs for example in Geoffroy's horseshoe bat.

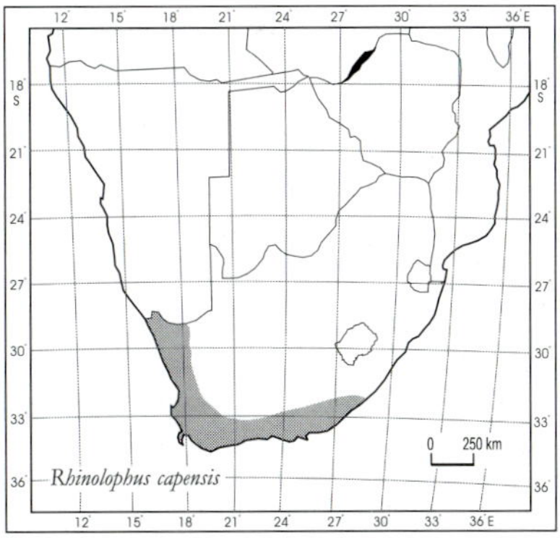

DAY ROOST

Caves and disused mines.

CALL

CF-FM: 88–90 kHz, duration and structure unknown.

DISTRIBUTION AND STATUS

Endemic to South Africa, and listed as Vulnerable in the *1996 IUCN Red List of Threatened Animals*, probably due to it being restricted to relatively few suitable underground roosts in the Western Cape and Eastern Cape provinces of South Africa.

	Males			Females		
FA	Mean 49	Range 48–52	*n* 9	Mean 47	Range –	*n* 1

Source: Skinner and Smithers (1990); Taylor (unpublished)

Geoffroy's horseshoe bat *Rhinolophus clivosus* (Cretzschmar 1828)

Second largest southern African horseshoe bat, FA 49–57 mm; connecting process blunt; anterior premolar crowded outside toothrow, not visible with hand lens, canine and posterior premolar touching; colour greyish or reddish-brown, paler below, wing and tail membranes dark; CF frequency 94 kHz.

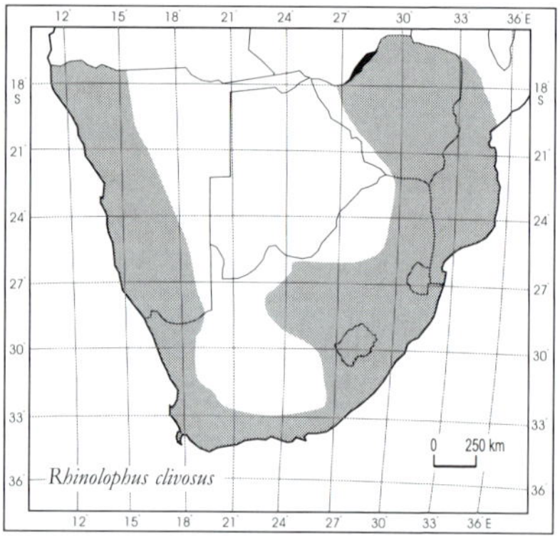

Southern African distribution only

HABITAT

Typically savanna woodland, but also deserts in the drier west, as well as grassland habitats in the Drakensberg, wherever suitable rocky areas are available for roosting.

HABITS

Colonies of up to 10 000 recorded in caves such as De Hoop and Koegelbeen Cave in the Western Cape and Northern Cape provinces respectively.

FEEDING

Like the previous species, they are slow aerial 'clutter foragers' as well as 'perch-hunters' and possibly gleaners. Their diet comprises predominantly beetles.

REPRODUCTION

Gestation is 3.5 months. Females are capable of breeding in their first year, and give birth once a year, in December, to a single young. Mating occurs before winter hibernation, and the sperm of the male is stored in the female reproductive tract during winter. Ovulation and fertilisation of the ovum follow in spring.

DAY ROOST

Caves, disused mines and similar cavities.

NIGHT ROOST

Trees, eaves and verandahs of buildings.

CALL

CF component of 94 kHz with FM sweeps down to 74 kHz, and duration of approximately 20 ms.

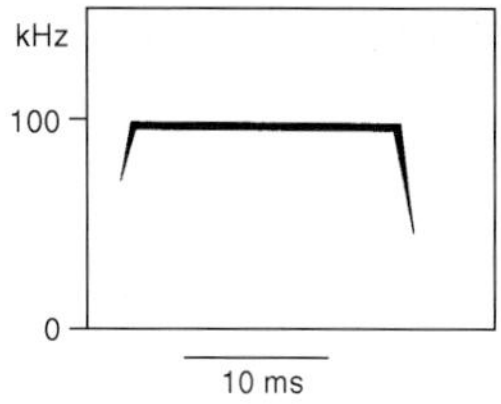

DISTRIBUTION AND STATUS

Fairly widespread and common throughout southern Africa, but absent from Botswana, eastern and northern Namibia, and much of the drier interior of the Northern Cape. North of southern Africa, they occur throughout the eastern and northern parts of Africa through to the Middle East, with a single record from West Africa.

	Males			Females		
	Mean	Range	*n*	Mean	Range	*n*
FA	53	49–56	38	53	51–57	30
Mass	16	13–20	25	17	12–25	26

Source: Rautenbach (1982); Taylor (1998b)

Darling's horseshoe bat *Rhinolophus darlingi* (K. Andersen 1905)

A small to medium-sized species (FA 43–49 mm, mass 8–10 g); similar in size to the Cape horseshoe bat (FA 48–52 mm), but distinguished from this species by having the anterior premolar crowded outside the toothrow, not visible with hand lens, canine and posterior premolar touching; colour drab grey to browner above, paler below; specimens from Namibia are paler and larger in size than specimens from elsewhere.

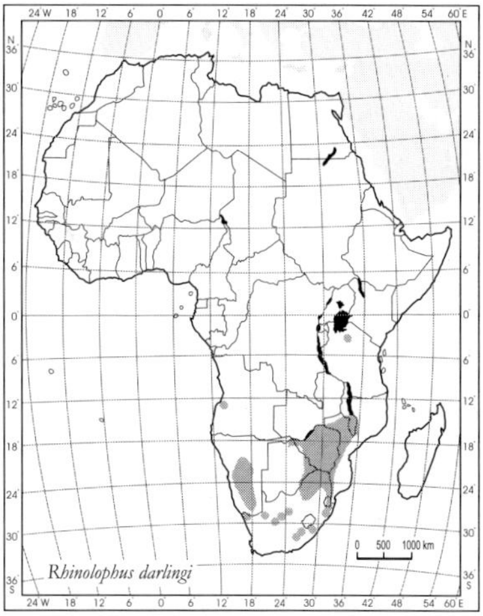

HABITAT

Savanna woodland, often associated with rocky or broken terrain.

HABITS

Not as gregarious as Geoffroy's horseshoe bat; colonies of up to a few dozen are known.

DAY ROOST

Caves, mines, fissures in rock outcrops.

CALL

Very long duration (35 ms), high frequency call comprising a CF component at 86 kHz, preceded and followed by short FM sweeps down to 80 kHz.

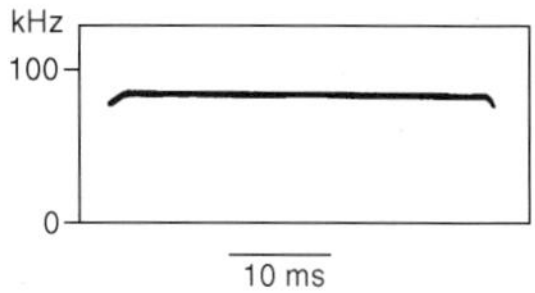

DISTRIBUTION AND STATUS

Zimbabwe, eastern Botswana, western Namibia, southern Mozambique, Swaziland, the Northern Province, Mpumalanga, Gauteng, KwaZulu-Natal, the Northern Cape and the Eastern Cape. North of southern Africa, they occur as far north as southern Malawi with isolated records from Angola and Tanzania.

	Males			Females		
	Mean	Range	*n*	Mean	Range	*n*
FA	45	43–48	16	47	43–49	30
Mass	8	7–10	18	9	7–14	32

Source: Rautenbach (1982); Smithers and Wilson (1979); Taylor (1998b)

Dent's horseshoe bat *Rhinolophus denti* (Thomas 1904)

Smallest of the horseshoe bats, FA 42–44 mm, mass 6 g; colour pale grey to pale cream above, with white bases to the longish dorsal hairs, white or off-white below, wing and tail membranes translucent brown with white edging; very similar in size to Swinny's horseshoe bat, and may represent the same species; for the time being their mutually exclusive habitats and ranges, at least in southern Africa (Dent's in arid, Swinny's in savanna), and differences in fur colour and noseleaf shape, separate the two species (see Chapter Five).

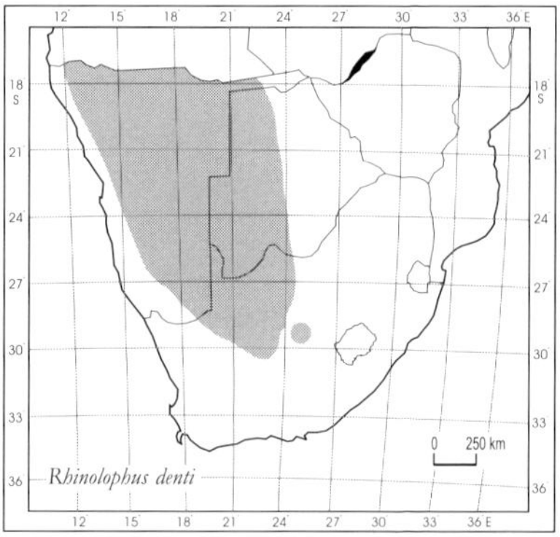

HABITAT

Arid habitats in the west of the subregion, wherever rocky outcrops or caves occur.

DAY ROOST

Caves, rocky caverns, open thatched huts.

CALL

Long-duration (15 ms) high frequency calls with CF component at 110 kHz, and FM component sweeping down to 82 kHz.

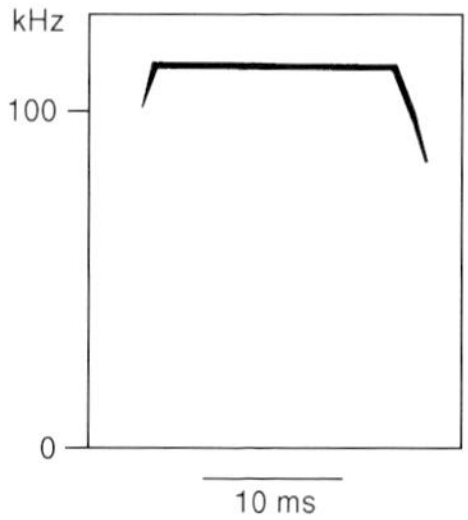

DISTRIBUTION AND STATUS

Apart from their reported occurrence from an isolated record from West Africa, restricted to Namibia, northwestern Botswana, the Free State, and the Northern Cape.

	Males			Females		
	Mean	Range	*n*	Mean	Range	*n*
FA	42	40–42	8	42	40–44	8
Mass	6	–	35	7	–	35

Source: Smithers (1971)

Rüppell's horseshoe bat *Rhinolophus fumigatus* (Rüppell 1842)

Width of horseshoe >9 mm, long hairs on face, and sella of noseleaf, as in R. hildebrandtii; but smaller than this species in size, FA 48–51 mm; CF frequency low, 55 kHz.

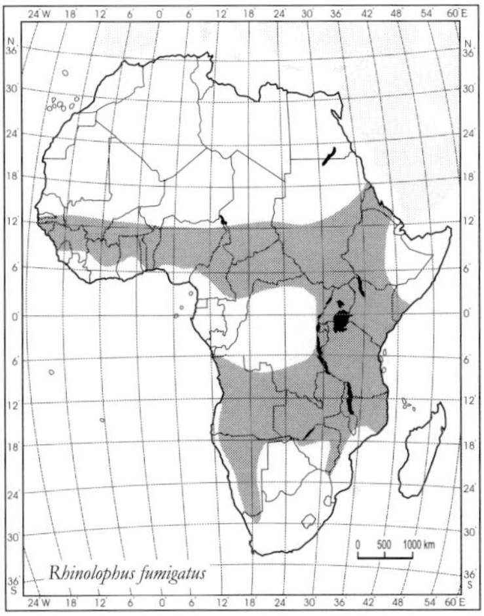

HABITAT

Open savanna woodland.

HABITS

Occur in small colonies of up to 12.

DAY ROOST

Caves, disused mines, piles of boulders, rock cavities.

CALL

Long-duration (15 ms) low frequency calls with CF component at 55 kHz, and FM component sweeping down to 50 kHz.

DISTRIBUTION AND STATUS

More widespread than Hildebrandt's horseshoe bat, its range extending throughout much of Namibia, excluding the Namib Desert.

	Irrespective of sex		
	Mean	Range	*n*
FA	50	48–51	8
Mass	16	12–20	5

Source: Smithers and Wilson (1979)

Hildebrandt's horseshoe bat *Rhinolophus hildebrandtii* (Peters 1878)

Largest horseshoe bat occurring in the subregion, FA 62–67 mm; horseshoe broad, more than 9 mm at its greatest width, and face and sella of noseleaf covered with long hairs; greyish-brown above, only slightly paler below; wing and tail membranes translucent brown; ears large and prominent, sharply pointed; CF frequency very low, 37–46 kHz.

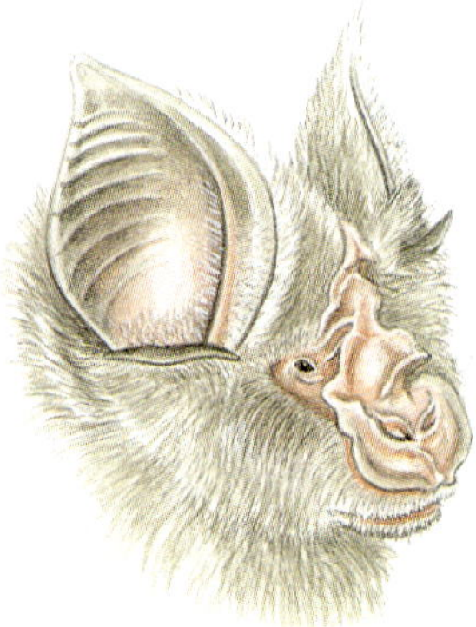

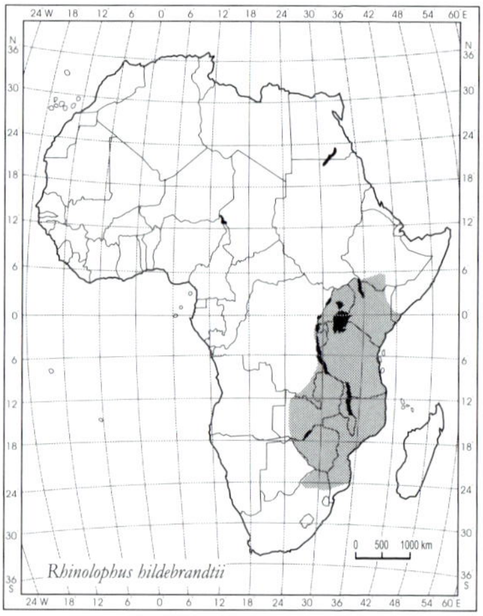

HABITAT

Savanna woodland.

HABITS

They are colonial, occurring in groups of up to hundreds, manoevrable flyers, able to forage in dense clutter.

FEEDING

They are slow aerial 'clutter foragers', which are capable of feeding by 'perch-hunting' and possibly also by gleaning. Diet includes beetles and moths; also some flies.

DAY ROOST

Caves, disused mines, tree hollows, aardvark burrows, roofs of houses.

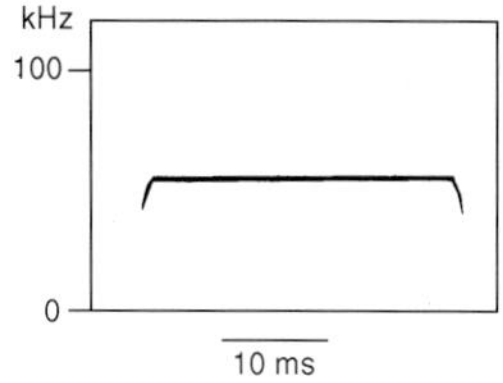

CALL
Although some variability has been reported, they have long-duration (15 ms), relatively low frequency calls comprising a CF component at 40 kHz, with FM component sweeping down to approximately 24 kHz.

DISTRIBUTION AND STATUS
Widely distributed in the northeast, especially Zimbabwe.

	Males			Females		
	Mean	Range	*n*	Mean	Range	*n*
FA	64	61–69	20	65	62–67	12
Mass	26	20–32	21	29	24–34	16

Sources: Rautenbach (1982); Smithers and Wilson (1979)

Lander's horseshoe bat *Rhinolophus landeri* (Martin 1838) (Plate 15)

One of the smaller horseshoe bats; similar to the peak-saddle horseshoe bat in overall size and in the pointed shape of the connecting process of the noseleaf, but distinguished from the latter species by the presence of underarm (axillary) tufts in the males, and on the basis of the proportions of the fourth digit (see Chapter Five); colour is brown above, lighter brown below, occasionally occurring in a reddish phase.

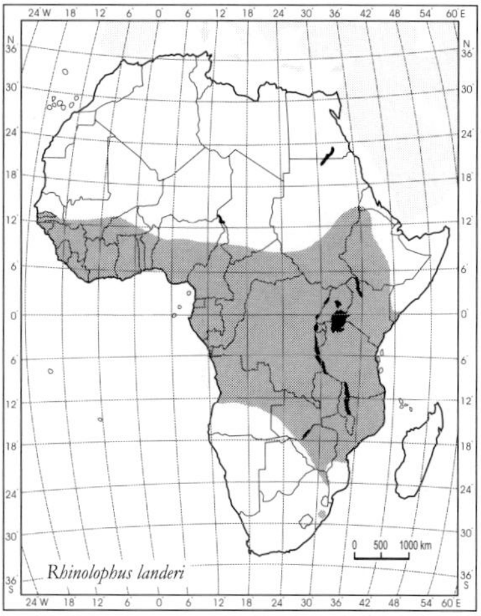

HABITAT
Wide habitat tolerance, from tropical forests to dry woodlands.

HABITS
They occur in small groups of less than 12.

FEEDING
Slow aerial 'clutter foragers' which also 'perch-hunt' and may glean their prey from surfaces. They appear to feed selectively on noctuid moths.

DAY ROOST
Hollow roosting: caves.

CALL
Medium duration (15 ms) CF-FM call with a maximum frequency of 110 kHz, sweeping down to 105 kHz.

DISTRIBUTION

Zimbabwe, Mozambique, and Mpumalanga, with an isolated record from KwaZulu-Natal.

	Males		
	Mean	Range	*n*
FA	42	41–43	18
Mass	7	6–9	18

Source: Skinner and Smithers (1990)

Bushveld horseshoe bat *Rhinolophus simulator* (K. Andersen 1904)

Recognisable by its relatively small size, and the longer ears (see Chapter Five), distinguishing it from the similar-sized Darling's horseshoe bat; echolocation call also distinct from Darling's horseshoe bat, with CF frequency of 83 kHz (86 kHz in Darling's). Observations in South Africa (but not Zimbabwe) show them to have a pale, whitish belly which is somewhat paler than the dorsal colour.

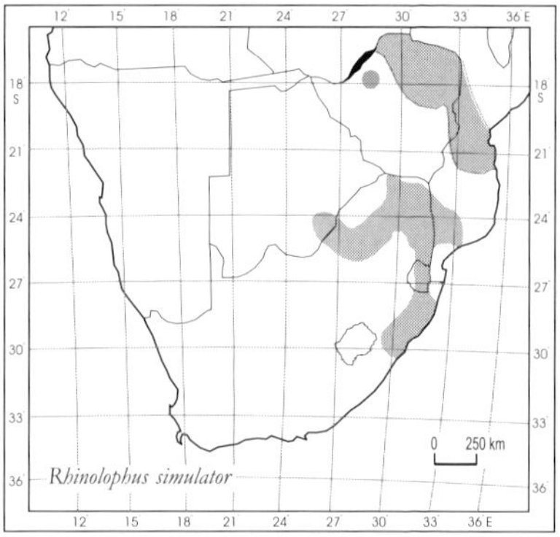

Southern African distribution only

HABITAT

Savanna woodland and riverine woodland.

HABITS

Local migrations of females, but not males, have been noted in KwaZulu-Natal and Zimbabwe.

DAY ROOST

Mines and caves.

CALL

The long-duration (18 ms) call comprises a high frequency CF component at 83 kHz, followed by an FM component sweeping down to 70 kHz. The fundamental component of the FM sweep is represented by a steep-FM component at 38–30 kHz.

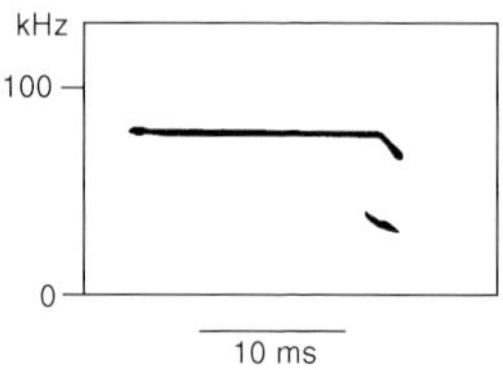

DISTRIBUTION AND STATUS

Scattered distribution in the eastern parts of the subregion extending along the coast as far south as the southern border of KwaZulu-Natal.

	Males			Females		
	Mean	Range	*n*	Mean	Range	*n*
FA	44	42–46	26	46	43–48	41
Mass	9	7–13	20	10	8–13	51

Sources: Rautenbach (1982); Seamark (pers. comm.); Taylor (1998)

Swinny's horseshoe bat *Rhinolophus swinnyi* (Gough 1908)

Its small size (FA 43–44 mm, mass 8 g) and its high CF frequency of 110 kHz, distinguishes it from the other small species, bushveld and Darling's horseshoe bats, whereas its colour, noseleaf and distribution (see Chapter Five) distinguish it from the similar-sized Dent's horseshoe bat.

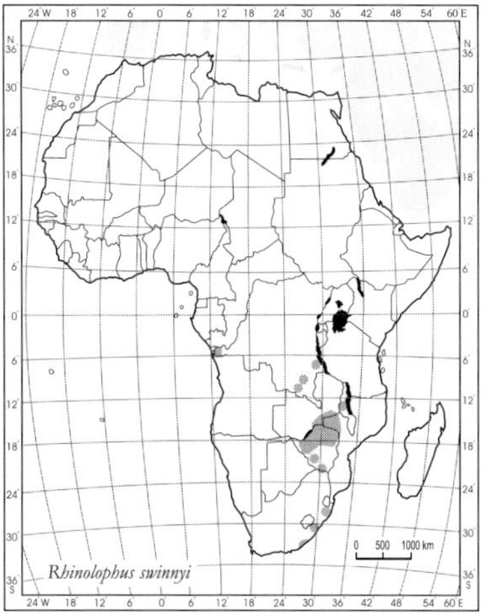

DAY ROOST

Caves, rocky caverns, open thatched huts.

CALL

Long-duration (15 ms) call with high CF component at 110 kHz, with FM component sweeping down to 82 kHz.

DISTRIBUTION AND STATUS

Fairly common in Zimbabwe, few records available further south, and not collected in the Eastern Cape for over half a century.

	Irrespective of sex		
	Mean	Range	*n*
FA	43	43–44	5
Mass	7.6	7.4–7.7	5

Source: Skinner and Smithers (1990)

TRIDENT AND LEAF-NOSED BATS – HIPPOSIDERIDAE

This family can be distinguished from Rhinolophidae on the basis of the simpler structure of the noseleaf, which lacks the single, triangular, erect process (lancet) typical of rhinolophids. The noseleaf has a simple elliptical shape in the genus *Hipposideros,* while in the other two genera of hipposiderids, *Cloeotis* and *Triaenops*, the noseleaf has a characteristic three-pronged process. Like horseshoe bats, hipposiderids use CF-FM echolocation calls, but generally of shorter duration and higher frequency, making bat detectors an invaluable tool for identifying these bats on the wing in caves. As also found in horseshoe bats, the CF frequency is always the dominant frequency, and it is thought that this component usually represents a second harmonic, with the fundamental component being suppressed to some degree.

Cloeotis (Thomas 1901)

Short-eared trident bat *Cloeotis percivali* (Thomas 1901) (Plate 16)

The three-pronged, trident-like process on top of the noseleaf is possessed only by this species and the rarer and larger Persian leaf-nosed bat; size tiny, mass of 3–6 g, total length of 62–76 mm, FA 33–39 mm; colour slate-grey to buffy brown above, whitish-yellow below, with soft, long fur, face yellowish; ears small, almost circular, forward facing and close to the eyes; echolocation call distinctive, with very high fundamental CF frequency at 104 kHz and first harmonic at 208 kHz.

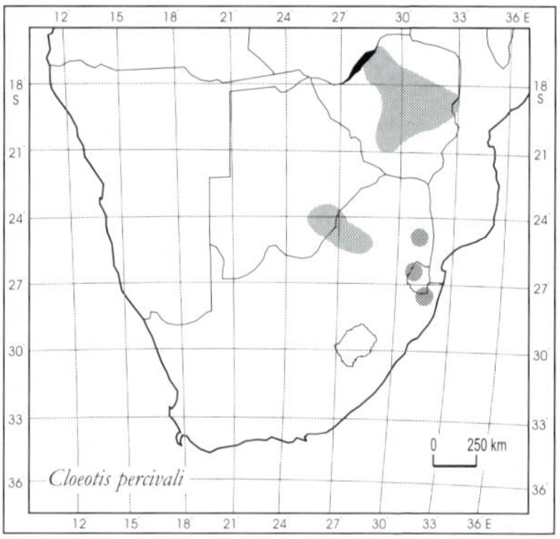

Southern African distribution only

HABITAT

Savanna woodland, specific details lacking.

HABITS

Colonies of up to hundreds are known.

FEEDING

They are slow, aerial 'clutter foragers', feeding exclusively on moths.

REPRODUCTION

Females carrying their newborn young have been found in December.

DAY ROOST

Hollow-roosting: caves, mine shafts, irrigation tunnels, and cavities in dam walls.

NIGHT ROOST

Individuals were accidentally killed coming into a ranger's house at Mkuzi Game Reserve, suggesting that they enter houses when foraging, presumably to use as night roosts.

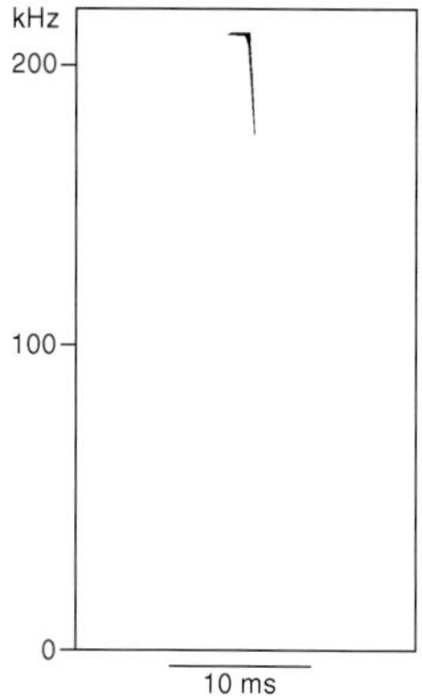

CALL

They are highly sophisticated echolocators, adapted for a diet specialising in moths. This species has the echolocation call with the highest known frequency for any bat, with short (3 ms) CF components at 104 kHz (fundamental) and 208 kHz (first harmonic). This extremely high frequency is an adaptation to avoid the sensitivity range of hearing ('tympanate') moths which comprise their diet, preventing the moths from hearing their echolocation calls and taking avoidance measures.

DISTRIBUTION AND STATUS

Apart from some five isolated extralimital records from coastal Kenya, Zambia and the northern Tete province of Mozambique, the known range of this species is largely confined to southern Africa, where its distribution is patchy. They are known from some six localities in Zimbabwe, two localities from southeastern Botswana, and single localities from Swaziland, the North-West Province, the Northern Province, and KwaZulu-Natal. Since October 1996, listed in the Endangered Mammal category under the 1974 Natal Conservation Ordinance (Schedule 6), and listed as Specially Protected in the new KwaZulu-Natal Conservation Ordinance of 1999. Listed as Lower Risk (Near Threatened) in the *1996 IUCN Red List of Threatened Animals*.

	Males			Females		
	Mean	Range	*n*	Mean	Range	*n*
FA	34	33–39	25	35	33–38	18
Mass	4.5	3.5–6.0	16	4.0	3.4–4.3	13

Sources: Rautenbach (1982); Smithers and Wilson (1979); Taylor (1998b)

Leaf-nosed bats *Hipposideros* (Gray 1831)

Two species occur. Commerson's leaf-nosed bat (*H. commersoni*) occurs in the Pafuri area of the Kruger National Park in South Africa, as well as in Zimbabwe, Mozambique, Namibia, and Botswana, while Sundevall's horseshoe bat (*H. caffer*) is widely distributed throughout the eastern and northern parts of southern Africa.

Members of the genus are immediately recognised by the shape of the horseshoe described above. The two species are very easy to tell apart on size and appearance, Commerson's leaf-nosed bat being 10 times heavier than Sundevall's leaf-nosed bat, with a forearm which is over twice as long (see next page).

Sundevall's leaf-nosed bat *Hipposideros caffer* (Sundevall 1846)

A small species, with a FA of 45–50 mm, mass of 7–9 g; unlike Commerson's leaf-nosed bat, the fur is long and fluffy, not short; colour is variable, typically greyish-brown, frequently showing a bright reddish-yellow colour phase; ears are broader than the previous species, with pointed tips; pair of false nipples, right more developed than left, are present in the anal region in females.

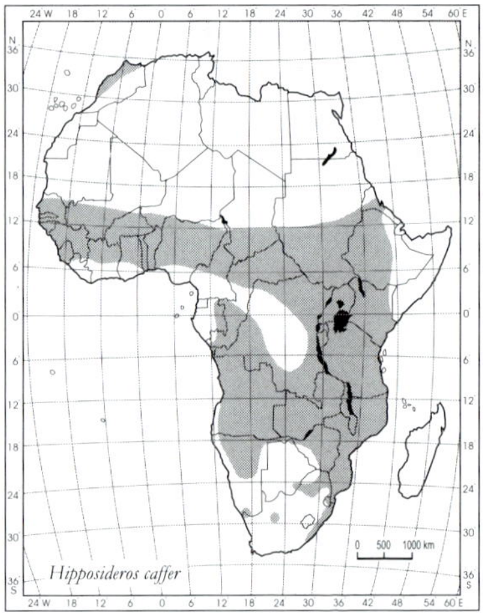

Southern African distribution only

HABITAT

Savanna woodland, usually associated with rivers and other water sources.

HABITS

Colonies vary in size from small groups to groups of hundreds. They roost hanging from the hind claws from the roofs and walls of caves, with individuals spaced at a distance of a few centimetres from one another.

FEEDING

They are slow aerial 'clutter foragers', possibly also capable of gleaning. Based on a study at Mkuzi Game Reserve, their diet consists mainly of moths, with fewer numbers of lacewings and beetles. Other studies show lacewings to be either absent, or a dominant item of the diet, indicating some flexibility in the diet.

REPRODUCTION

Gestation is 220 days, being delayed due to the period of retarded embryonic growth which occurs during winter hibernation. Females gives birth to a single offspring once a year early in December.

DAY ROOST

Very variable, from caves, mines and tunnels, to road culverts, rock fissures, and the roofs of houses.

NIGHT ROOST

Eaves of houses, under thatched rondavels, in disused buildings.

CALL

Short-duration (7 ms) call with CF component at 138 kHz (recorded in Zimbabwe) or 144 kHz (recorded in KwaZulu-Natal), followed by an FM sweep down to 105 kHz (Zimbabwe) or 120 kHz (KwaZulu-Natal).

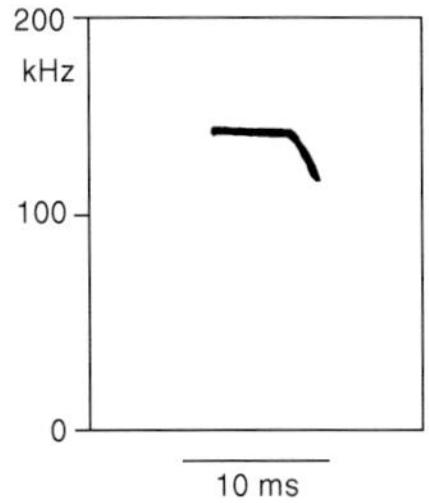

DISTRIBUTION AND STATUS

Common and widespread throughout the savanna biome in the northern and northeastern parts of the subregion, with isolated records from the arid lower reaches of the Orange River in the Northern Province and southwest Namibia.

	Males			Females		
	Mean	Range	*n*	Mean	Range	*n*
FA	48	45–50	48	47	42–50	42
Mass	8	7–11	42	8	6–15	38

Sources: Rautenbach (1982); Smithers and Wilson (1979); Taylor (1998b)

Commerson's leaf-nosed bat *Hipposideros commersoni* (E. Geoffroy1813) (Plate 17)

This is the largest insect-eating bat in the subregion, showing pronounced sexual dimorphism in size and colour; wing span almost 600 mm, mass 100–130 g in males (with breeding males reaching up to 180 g), and 70–80 g for females (with breeding females reaching 120 g), FA 100–108 mm in males, 90–101 mm in females; in males, the fur is short and sparse towards the rump and hindquarters, fawn coloured above, and light tawny below, with white flanks; in females fur is thicker and longer and a richer reddish-brown compared with more greyish colour of males; ears widely spaced and characteristically pointed; males have a well-developed frontal gland above the noseleaf, absent or poorly developed in females and non-breeding males.

HABITAT

Predominantly savanna woodland.

HABITS

Cave colonies of up to tens of thousands have been recorded. At Mabura cave in central Zimbabwe, the total population was conservatively estimated to be 100 000 when breeding females were in residence.

FEEDING
They are slow aerial 'moderate clutter' feeders, which may be capable of 'perch-hunting' and gleaning. Diet consists predominantly of large beetles.

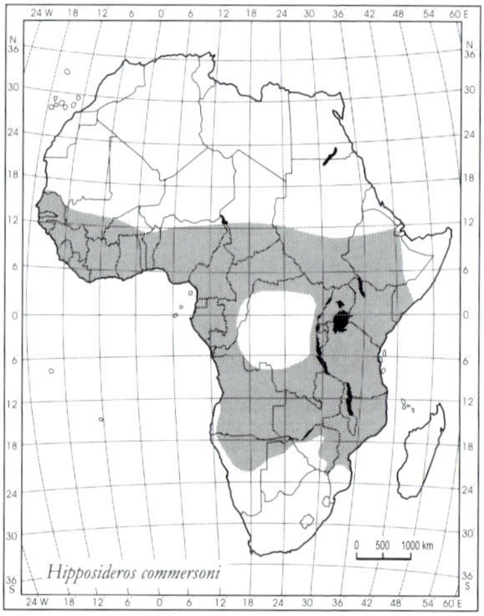

REPRODUCTION
Gestation lasts four months. Females give birth to a single offspring once a year. There does not appear to be any reproductive delay phenomenon.

DAY ROOST
Predominantly caves, although solitary individuals are known to roost within tree canopies or under the eaves of buildings.

NIGHT ROOST
An individual was observed using a tree as a night roost, on the fringe of the Okavango Delta in Botswana.

CALL
The medium-duration (12 ms) call comprises a 62 kHz CF component with FM sweep down to 55 kHz.

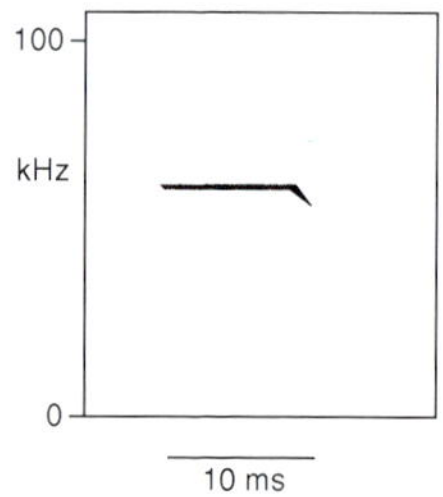

DISTRIBUTION AND STATUS
Northern Namibia, northern Botswana, Zimbabwe, Mozambique and the Pafuri area of the Kruger National Park. Few large colonies of this species are known, e.g. only three roosts are known in Zimbabwe. In Zimbabwe

and Angola, they are threatened by guano mining and harvesting of the bats for food and to extract fat to make oil (Cotterill, pers. comm.).

	Males			Females		
	Mean	Range	*n*	Mean	Range	*n*
FA	101	93–105	5	95	93–102	6
Mass	116	115–116	2	126	120–132	2

Source: Smithers and Wilson (1979)

Triaenops (Dobson 1871)

Persian leaf-nosed bat *Triaenops persicus* (Dobson 1871) (Plate 18)

Like the short-eared trident bat, it has a noseleaf possessing a three pronged trident; however, it is much larger in size, FA 48–54 mm, as opposed to 33–39 mm; colour yellowish-brown, underparts lighter, wing and tail membrane dark brown.

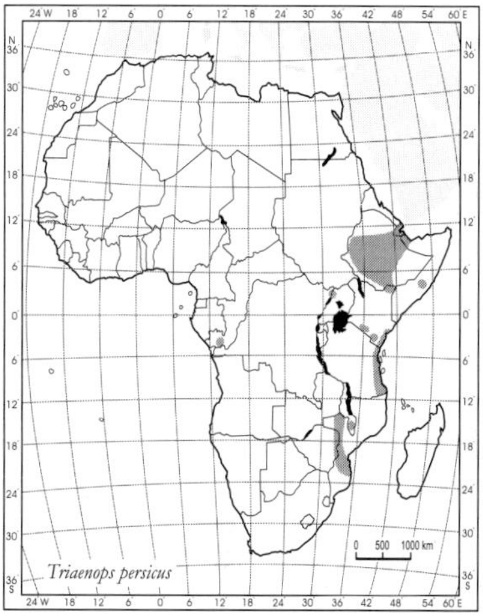

HABITS

They have been found singly, or in groups of hundreds, in caves.

DAY ROOST

Caves.

DISTRIBUTION AND STATUS

The Persian leaf-nosed bat is known from isolated records in Zimbabwe and Mozambique but is not found in South Africa, Botswana, or Namibia.

	Males		
	Mean	Range	*n*
FA	–	48–54	–

Source: Hayman and Hill (1971)

Section Three

Bat-watching

Chapter Seven

Bat groups and conservation

So you want to be a bat-watcher? The best way to get started is to join your nearest bat interest group or start your own. Of course you can also do bat-watching on your own.

There are currently three local bat conservation groups in southern Africa: the Durban Bat Interest Group (DBIG), Gauteng Bat Interest Group (GBIG), and Cape Bat Action Team (Cape BAT), as well as 11 centres represented by active bat-workers, and at least three speleological (caving) societies. Full contact details of all of these are provided under Important contacts. Most of the people and groups listed would be willing to hear from persons interested in joining a bat interest group, starting their own, or just learning more about bats, as well as from persons who may have dead bats requiring identification, e.g. those killed by cats or dogs, skeletal remains in owl pellets, or mummified corpses found in caves or attics. There is always the distinct possibility of coming across an important new species distribution record for your region, province or country.

If there are no bat interest groups near you, and you are a fanatical, frustrated bat-watcher, write to, phone, fax, or e-mail the nearest group to you to find out how to get the necessary training and literature to start up a new group.

OVERVIEW OF BAT CONSERVATION

Declines in bat populations have been documented worldwide. In many parts of the world, historical accounts described night skies filled with bats, where this is no longer the case. The greater horseshoe bat of Europe has become extinct in many areas of Britain where it was once common. Early explorers to Indonesia described colonies of flying foxes numbering millions, and which have now been reduced by hunting to a few thousand. Several of North America's largest cave colonies have suffered declines of up to 99.9%.

In recent years there has been a rapid increase in scientific and public interest in bat conservation. Nowhere has this been as dramatic as in Britain, where the Wildlife and Countryside Act of 1981–1982 (which made it illegal to kill or harm bats) stimulated the formation of numerous amateur bat groups to assist with enforcing the Act. From some 23 amateur bat groups in 1984, 90 groups with 2 000 members were known in 1992. The Nature Conservancy Council (now known as English Nature) issued 350 licences to bat-workers in 1992, compared with just a few in 1982. During the same period, the number of public enquiries about bats, registered by English Nature, rose sharply to an eventual peak of about 3 500 per year. About 80% of public enquiries were followed up by personal visits by a nature conservation official or an amateur bat-worker, providing a valuable source of biological information. From 400 known roosts in 1972, 14 000 building roosts and 1 000 underground hibernation roosts were known in 1992.

South Africa has 56 recorded species of bats, most of which are poorly known. Two species, the giant free-tailed bat and Natal free-tailed bat, have been recorded only from very old type-specimens. Microchiroptera are listed as Protected Wild Animals under Schedule 2 of the old Cape Provincial Ordinance No. 19 of 1974. In 1996, two species of bats were listed as Endangered Mammals in KwaZulu-Natal (Schedule 6 of the Natal Provincial Ordinance No. 15 of 1974). The revised KwaZulu-Natal Ordinance of 1999 recognises several species of bats as either Specially Protected or Protected. Elsewhere in southern Africa bats are not formally protected.

In the wider subregion of southern Africa (south of the Zambezi and Cunene Rivers), we don't yet know how bats are faring. Of the 74 species of bats found in southern Africa, 20 species of Microchiroptera (insectivorous bats) and two species of Megachiroptera (fruit-eating bats) are listed as Threatened in the *1996 IUCN Red List of Threatened Animals* – of these, nine are listed as either Critically Endangered, Endangered, or Vulnerable, meaning that they face a high risk of extinction in the wild in the immediate to medium-term future if the agents responsible for reducing their numbers are not curbed. The IUCN Red List category of each threatened southern African bat species was given before under the relevant species accounts in Section 2, and the list is summarised in the Table on the following page.

Because so few data are available for most bat species, Red List categories should not be regarded as definitive and final statements of conservation status, but as more data are gathered, the Red List categories should be continually updated (either by increasing or decreasing the level of threat) to best reflect the situation in nature. For example, recent collecting and surveys suggest that the following species are more common than suggested by their present category of Vulnerable, and probably not 'facing a high risk of extinction in the wild in the medium-term future': *Otomops martiensseni* (Richardson and Taylor 1995; Taylor 1998); *Tadarida lobata* (Cotterill 1996b); *Pipistrellus anchietai* (Kearney and Taylor 1997); and *Rhinolophus capensis* (Herselman and Norton 1981).

THREATS TO BATS

Major threats to the survival of bats worldwide include habitat destruction, widespread agricultural insecticide use, fumigation of house bats, over-hunting of both fruit bats and insect-eating bats for supplementing human diets (reported in southern Africa), or for sale as market delicacies and aphrodisiacs (particularly in Asia), and human disturbance of winter hibernating roosts and summer maternity roosts of cave-dwelling species. In both Central and South Americas vampire control programmes blast or poison hundreds of cave systems, destroying many species of bats and other animals apart from vampires. Behind all of the above threats lies the burgeoning human population and the negative perceptions which many people still harbour about bats.

In southern Africa, very few data are available on the status of bat populations and the nature and magnitude of threats to their continued survival. Some factors which may pose a threat to the survival of bats are listed below. Studies documenting the effect of such factors on bat populations and communities are urgently required.

Table of southern African bat species listed as Threatened in the *1996 IUCN Red List of Threatened Animals.* Detailed criteria for the following categories can be found on the Internet at **http://iucn.org/themes/ssc/redlists/criteria.htm**

LR (NT) = Lower Risk (Near Threatened): A taxon is Lower Risk (LR) when it has been evaluated and does not satisfy the criteria for any of the categories Critically Endangered, Endangered, or Vulnerable. Taxa included in the Lower Risk category can be separated into three subcategories, including Near Threatened (NT): taxa which do not qualify for Conservation Dependent, but which are close to qualifying for Vulnerable.

VU = Vulnerable: A taxon is Vulnerable when it is not Critically Endangered or Endangered but is facing a high risk of extinction in the wild in the medium-term future.

EN= Endangered: A taxon is Endangered when it is not Critically Endangered but is facing a very high risk of extinction in the wild in the near future.

CR = Critically endangered: A taxon is Critically Endangered when it is facing an extremely high risk of extinction in the wild in the immediate future.

Species	Category of threat
Angolan epauletted fruit bat *Epomophorus angolensis*	LR (NT)
Bergmans's small fruit bat *Myonycteris relicta*	VU
Pale free-tailed bat *Chaerephon chapini*	LR (NT)
Natal free-tailed bat *Mormopterus acetabulosus*	VU
Large-eared free-tailed bat *Otomops martiensseni*	VU
Madagascar large free-tailed bat *Tadarida fulminans*	LR (NT)
Big-eared free-tailed bat *Tadarida lobata*	VU
Giant African free-tailed bat *Tadarida ventralis*	LR (NT)
Lesser long-fingered bat *Miniopterus fraterculus*	LR (NT)
Schreibers's long-fingered bat *Miniopterus schreibersii*	LR (NT)
Botswana long-eared bat *Laephotis botswanae*	LR (NT)
Namib long-eared bat *Laephotis namibensis*	EN
De Winton's long-eared bat *Laephotis wintoni*	LR (NT)
Lesueur's wing-gland bat *Myotis lesueuri*	VU
Angola wing-gland bat *Myotis seabrai*	VU
Anchieta's bat *Pipistrellus anchietai*	VU
Thomas's house bat *Scotoecus albofuscus*	LR (NT)
Giant yellow house bat *Scotophilus nigrita*	LR (NT)
Wood's slit-faced bat *Nycteris woodi*	LR (NT)
Cape horseshoe bat *Rhinolophus capensis*	VU
Peak-saddle horseshoe bat *Rhinolophus blasii*	LR (NT)
Short-eared trident bat *Cloeotis percivali*	LR (NT)

Reliable witnesses have reported instances where cave bat colonies are being harvested for food by local communities in Zimbabwe and South Africa, and there are concerns that this may not be sustainable. Human pressure on a limited number of cave roosts is growing constantly. Previously important bat roosts like the Cango Caves, now a major tourist site, have not harboured bats for many years due to human use. In urban areas like Durban and Pietermaritzburg vandalism is common, where for example satanic groups burn fires and hold sacrifices in caves used by bats. Very few cave roosts fall in protected areas, making them vulnerable to uncontrolled disturbance. Different species of bats have specific microclimate requirements for maternity and hibernation, and a single cave or tunnel may function as the only maternity roost for a species for a very large surrounding area. Destruction of this one cave would impact severely on the population of the species, even causing local extinction.

Agricultural monocultures in different parts of the subregion, along with widespread use of agricultural pesticides must be having an adverse effect on bats, although this has yet to be quantified. This is particularly worrying since bats are top predators and the effects of chemicals such as DDT (which is still used in some areas) are concentrated in biological tissues as one moves up the food pyramid. Atmospheric pollutants resulting from industrial activities, such as the concentration of thermal power stations in regions of Gauteng and Mpumalanga, are likewise potentially harmful to bats and their insect prey. Similarly, widespread use of poisons such as methyl bromide and lindane in fumigating houses against wood borer and other such pests can have a devastating effect on house-roosting bats, particularly for example in the large-eared free-tailed bat, *Otomops martiensseni*, which occurs mostly in houses in the Durban area (see Chapter Nine).

A range of land uses degrade habitat quality which can adversely effect bat populations. A recent study in Zimbabwe showed that fewer bats occupied woodland habitats converted to cropland and damaged by elephants, compared with intact woodlands.

Research is urgently required to document conservation status and monitor long-term population changes, particularly for those southern African species listed as Threatened in the *1996 IUCN Red List of Threatened Animals*, and to investigate the nature and magnitude of threats to the survival of bats in southern Africa. Further recommendations from recent action plans drafted by the Chiroptera Specialist Group of the International Union for Conservation of Nature's Species Survival Commission (IUCN/SSC) include compiling an inventory of underground bat roost sites, with specific management plans for the more important sites, compiling a regional action plan (e.g. for southern Africa), forcing better legislation for protection of bats, liaising closely with IUCN/SSC and other international and local conservation non-government organisations (NGOs) and, above all, educating the public and decision-makers, including compiling a catalogue of currently available educational material on bats.

BAT GROUPS IN SOUTH AFRICA

Bat interest groups were formed in KwaZulu-Natal, Gauteng and the Western Cape in 1994, 1995 and 1999 respectively, with the primary aim of promoting greater public awareness of bats and bat conservation. The success of this approach has been gauged by a study based on statistics gathered for the Durban Bat Interest Group (DBIG) from 1994 to 1997 inclusive (Taylor 1999). Some of the results of this study are summarised below.

Inspired by the hugely successful British bat group movement, DBIG was officially launched on 26 February 1994, as a project of the Durban Natural Science Museum's Friends Society. DBIG's aims are to promote bats, bat conservation and bat research, principally in KwaZulu-Natal, through the training of volunteers, regular talks, meetings and outings for members – thus increasing public awareness of bats. Public awareness is promoted through advising and assisting the public and other organisations with bat-related queries and problems, giving public talks, participating in environmental exhibitions, and involving the media.

The number of public queries and contacts registered by DBIG increased over the past six years, from 106 in 1994 to some 250 in 1999. DBIG numbered some 110 members at the time of going to press. The upward trend in public interest has been encouraging and this has been continued throughout South Africa during 1999 and 2000, with the joint participation of bat groups and scientists from across the country in successful bat exhibitions at the Sasol Science Festival in Grahamstown (1999 and 2000), and two high profile visits to South Africa during 2000 from international bat conservationists and scientists, Dr Merlin D. Tuttle (Bat Conservation International) and Professor M.B. Fenton (York University, Ontario, Canada). South African bat groups currently participate in reciprocal membership exchanges with Bat Conservation International, Bat Conservation Trust of Britain, and Northern Ireland Bat Group.

Bat rescue

Many people come across young, injured or grounded bats, and wonder what to do with them. Care and caution should be exercised and such cases should be referred to the nearest bat interest group or animal rehabilitation organisation. DBIG receives some 60 such individual bats per year, of which two thirds are saved from certain death. About one third of all bats received are rehabilitated and subsequently released into the wild. Apart from the humane and educational benefits of this work, valuable scientific data have been gathered on the classification, reproduction, captive behaviour, and veterinary care of bats.

It must be stressed that bats should not be handled by the general public. Only trained experts in possession of nature conservation permits are permitted to keep bats in captivity. A comprehensive manual on Bat Care and Rehabilitation has been compiled by DBIG co-founder, E.J. Richardson, for *bona fide* rehabilitators or for those wishing to become further involved in bat care after appropriate training and supervision by their local bat interest group. The manual offers practical first aid advice to those faced with dealing with grounded or injured bats on a once-off or occasional basis, e.g. how to distinguish natural torpor from sickness, how to determine age and sex of a bat, and what and how to feed insectivorous bats, both babies and adults. For more information on bat care and handling, the manual mentioned here, or how to get involved in bat rehabilitation, contact one of the bat interest groups (see Important contacts).

Should you come across a stranded bat and have to handle it, make sure you use protective gloves or a towel. If the bat seems healthy, try hanging it up in a safe high spot, such as a window sill or tree. It may have just been stunned, in which case it should fly when it gets dark. Remember that many bats go into a kind of temporary hibernation (called torpor) during the day, especially during the winter months. This would make them very sluggish, but does not mean they are sick or dying. A baby bat that has lost its mother can often be successfully re-united with her, by placing it somewhere high and flat, near where you found it. The distress calls of the baby will attract the mother (or at least a surrogate mother) to swoop down and pick up the baby. Remember that not every little bat is a baby: many species of bats are very little as adults. Baby bats are not fully furred and tend to have very short stubby wings. Also, the ends of the finger bones of baby bats are un-ossified, and this shows up as a thin white line of cartilage when viewed against a concentrated light source (Figure 40).

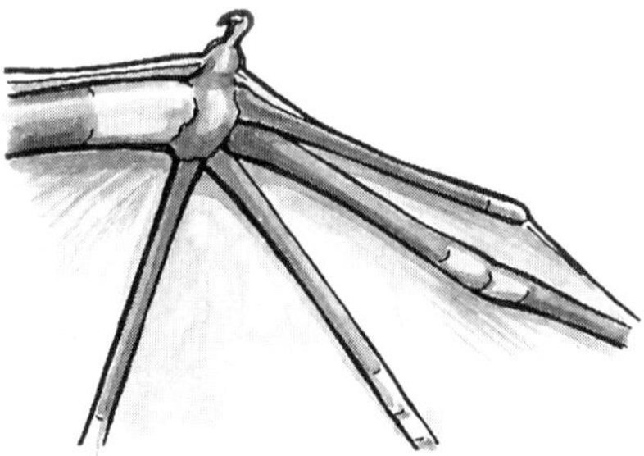

Figure 40 Juvenile horseshoe bat wing bones showing unfused epiphyses

Bat research and data collection

One of the most significant benefits resulting from DBIG's activities has been the stimulation of research and data collection efforts. A South African database of bat roosts has been compiled and currently registers over 600 entries. Two new species of bats, Rendall's serotine bat and Thomas's house bat, have been added to the list of South African species. Significant new biological data have been contributed for several species, including the tomb bat, large-eared fruit bat, Wahlberg's epauletted fruit bat, and short-eared trident bat.

STARTING YOUR OWN BAT GROUP

There seems to be a demand for more local bat interest groups in southern Africa. As is the case with bird clubs, such groups can be formed by enthusiastic volunteers who are not necessarily scientists, but supervision and training by scientists or trained bat-workers from existing bat groups is essential. A suggested plan for starting new bat groups is outlined below.

1) Advertise your intent to form a bat club in the local press, publicise dates for a slide show or talk on bats for interested persons, and also make the offer to visit any 'problem' roosts in the roof spaces of residential dwellings. This will be sure to get people's attention, but be prepared, the response may be overwhelming. If possible, get an interested reporter to write an article on bats, with a local flavour if possible (e.g. a rare bat found nowhere else).
2) Set a date and venue for the slide presentation or talk. Either invite an expert along to talk, or buy or borrow Bat Conservation International's (BCI's) superb slide series entitled 'Bats: Myth & Reality' (contact details for BCI on p. 186).
3) At the presentation ask interested persons to fill in application forms for your bat club. Be sure to insert a question enquiring who would be interested in attending a training course for bat-workers. Advertise a date in the near future for the envisaged training of the first batch of recruits.
4) Conduct the training course. Members of the existing bat interest groups would be very happy to assist (contact details on p. 184). Two full days, perhaps over a weekend, seems sufficient. If possible, organise a combined day visit to a nearby bat roost, or, alternatively, an evening bat walk with bat detectors (you will need to borrow these from one of the existing bat groups). Bat workers' manuals suited to southern African bats have been prepared by both DBIG and GBIG and can be obtained from these groups. After the training session, volunteer bat-worker recruits can be organised into teams to do practical follow-up work visiting roosts in houses or caves. I suggest that not everyone who does the course need go on to qualify as a bat-worker to do roost visits.
5) Have at least one combined outing to a known bat roost, where recruits can get a close-up look at real live bats.
6) Appoint reasonably qualified supervisors for each 'bat squad'. Each squad should have about four to 10 members. The supervisors will need to arrange at least three visits to bat roosts with their squad members (public calls in response to earlier publicity should provide a good source of addresses of people with nuisance bats in their roof space who would appreciate a visit).
7) When the squads have finished their practical training, set a simple test including a section which requires trainees to identify bat specimens from a key (these can be borrowed from your nearest natural science museum). Trainees who pass this get awarded a bat-worker's certificate and can therefore be used to follow up public enquiries, conduct roost visits, and give talks.
8) As soon as the club gets going, appoint a committee consisting of a chairman or newsletter editor, secretary or treasurer, and other portfolios such as programme organiser, fundraiser, and data recorder. Try to publish a short

newsletter at regular intervals to keep your members interested and informed.

9) Decide on a format for meetings and outings.
10) Get involved in your community through expositions and talks to schools and community groups, garden clubs, and environmental groups. To raise funds for worthy conservation, education, or research projects, get people interested in selling bat merchandise, such as bat houses, pencils, envelope labels, tin badges, brooches, coffee mugs, and T-shirts. A range of bat merchandising products, as well as educational brochures, can be obtained through the Durban Bat Interest Group or Gauteng Bat Interest Group. Also, stay in touch with other bat groups to see how you can contribute important information on the locations of local bat roosts, bat house projects, and so on.

Getting a global perspective

Bats worldwide are under pressure from human exploitation and ignorance, as well as habitat destruction. Yet bat conservation groups worldwide are beginning to make real progress with understanding and conserving bats. To stay abreast of these developments, bat-watchers are encouraged to join groups such as Bat Conservation International and the British Bat Conservation Trust (see Important contacts). For those with e-mail, it is well worth joining the Batline listserv, simply by sending an email to: **majordomo@lads.com** and typing in the message line:

'subscribe batline [your name]' (without the quotes, and insert your own name).

Chapter Eight

Echolocation and bat detectors

Bat detectors are to the bat-watcher what binoculars are to the bird twitcher. No keen bat enthusiast should be without one. In Europe, North America, and Australia, bat detectors are widely used for research, field surveys, or just for fun. For as little as R300, someone with no knowledge of bats or echolocation can use a bat detector, but far more valuable insights and enjoyment can be gained with a deeper knowledge of bat echolocation and the principles behind bat detectors. This chapter seeks to provide this background information in terms as simple as possible. Supplier contacts and tips on what bat detector to buy are also provided.

This chapter has borrowed ideas and illustrations (with permission) from an excellent introductory book (and accompanying CD) called *The bat detective* by Briggs and King (1998). For further technical details, readers are referred to an informative manual on bat echolocation by Catto (1994).

Figure 41 Cape horseshoe bat in pursuit of moth showing echoes (dashed lines) reflected back to the bat

WHAT IS ECHOLOCATION?

In 1793, Italian scientist, Lazarro Spallanzani first demonstrated that bats can 'see in the dark'. He put a bat and an owl in a candle-lit room and found that both could navigate in low light. When the candle was extinguished, the bat flew effortlessly in total darkness, but the owl collided with objects in its flight path. When Spallanzani interfered with the bat's hearing, it, too, became disoriented. Spallanzani concluded that bats used a 'sixth sense' to orient, but he was not satisfied that he had the final answer.

It was only 140 years later, in the 1930s, that American researcher Donald Griffin, using special microphones, showed that the little brown bats he studied produced sounds above the human range of hearing (i.e. ultrasonic). Griffin showed that the little brown bats used the echoes of the sounds they produced to locate objects in their path. It was Griffin who coined the word 'echolocation'. Donald Griffin's book, *Listening in the dark* (1958), has recently come back into print and is a must for bat enthusiasts.

Not all bats echolocate and some other animals also use this form of orientation. Dolphins, porpoises and toothed whales, some shrews and tenrecs, and oilbirds and several species of swiftlets use echolocation.

Echolocation is mostly restricted to microbats. Fruit-eating megabats rely mainly on sight and smell to find food, and only one, the Egyptian fruit bat (*Rousettus egyptiacus*) uses echolocation to orient itself in the pitch dark caves where it roosts.

The elaborate facial noseleaves of some microchiropteran bats are thought to be associated with echolocation, acting like a megaphone to concentrate and focus sound. Bats with noseleaves always emit their echolocation calls through the nose, as opposed to many other bats (such as vespertilionids) which echolocate through their mouths, and thus fly with their mouths open. Both nasal emitters and oral emitters can chew their food and vocalise at the same time. All microbats produce their echolocation signals in their voice box (larynx), while Egyptian fruit bats echolocate using clicks produced by clicking their tongues.

Not all sounds produced by bats are echolocation calls. Bats produce many social calls which may or may not be audible to the human ear, for example, at their roost, to communicate among individuals or between mothers and young, or, in flight, to defend their foraging territories. Male pipistrelles (*Pipistrellus pipistrellus*) emit an elaborate ultrasonic courtship song associated with a particular courtship flight pattern.

An echolocating bat registers each outgoing sound pulse and compares the original sounds with the returning echoes. The time lapsed between generating the outgoing sound pulse and receiving the incoming echo provides an accurate measure of a target's distance from the bat. Changes in the intensity and frequency (pitch) of the outgoing sound provide information about the nature of the target, such as its size, shape, surface structure, and speed of movement.

Incoming sounds, including echoes, are received by the bat's external ears, or pinnae, before they are channelled into the rest of a bat's hearing system. The huge variety in bat ear shape reflects differences in the hearing abilities of different species, and speaks of the great diversity of bat echolocation and foraging behaviours.

WING SHAPE AND FLIGHT BEHAVIOUR

Since they are both closely tied to the way that bats forage, wing shape and flight behaviour should be considered together with echolocation calls, when it comes to field identification of bats.

Some terms are useful when it comes to describing wing shape. Wing loading is defined as the bat's body weight carried in flight, divided by its wing area. Bats with low wing loading can fly at slower speeds without stalling. Aspect ratio is defined as the square of the wingspan, divided by the wing area. Long, narrow wings have a high aspect ratio, while short, broad wings have a low aspect ratio.

Bats with high aspect ratios and high wing loadings (such as most free-tailed bats, Figure 42) will have faster flight, a faster stalling speed, and be less capable of manoeuvring in confined spaces. They will tend to fly fast to their feeding areas and forage in more open spaces. On the other hand, bats with low aspect ratios and low wing loadings (e.g. horseshoe bats and slit-faced bats, Figure 43) will have a slower, more 'energetically expensive' flight, but will be able to fly slowly in cluttered habitats, to hover, make tight turns in slow flight, and glean insects from branches. They will also be able to carry large insect prey back to their feeding stations.

It is also useful to understand the function of different parts of the bat's wing (Figure 44). The arm wing (plagiopatagium) and tail membrane (uropatagium) are primarily load bearing. The hand wing (dactylopatagium) produces the thrust for forward movement. The leading edge, in front of the arm bones (propatagium),

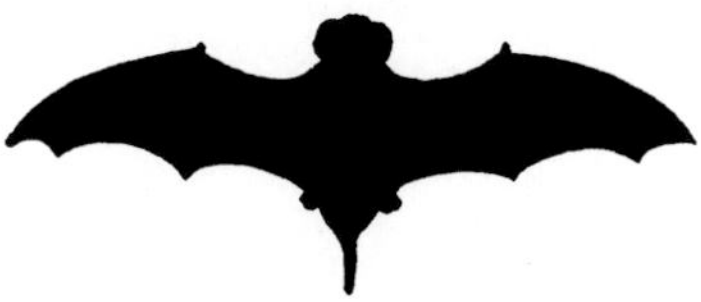

Figure 42 Outline of little free-tailed bat, a species with high aspect ratio wings

Figure 43 Outline of Geoffroy's horseshoe bat, a species with low aspect ratio wings

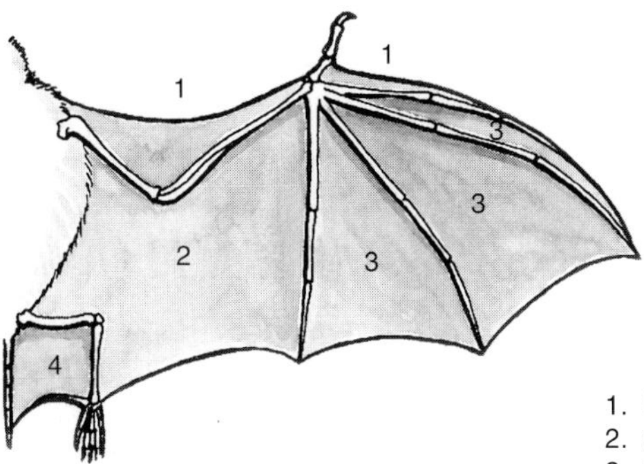

Figure 44 Wing parts

1. leading edge (propatagium)
2. arm wing (plagiopatagium)
3. hand wing (dactylopatagium)
4. tail membrane (uropatagium)

determines the angle of attack of the wing. Fast-flying bats have a small angle of attack (hence a small propatagium), while slower, manoeuvrable flyers, such as horseshoe bats, have a larger propatagium, therefore a steeper angle of attack, and a greater ability to increase the camber (front to rear curvature) of their wings.

BACK TO BASICS – PHYSICS OF SOUND

Before going any further, it is important to clarify some basic terms which are necessary to further understand how bat detectors work, and why and how bats vary in their echolocation calls.

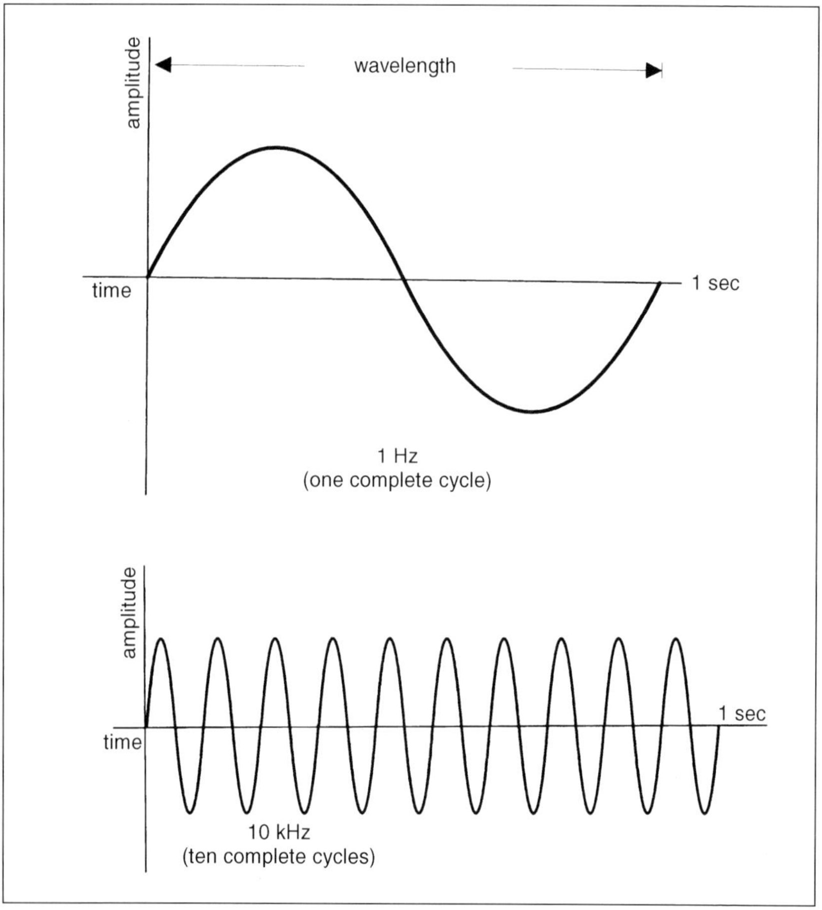

Figure 45 Diagram illustrating the relationship between wavelength and frequency (measured in Hertz)

Wavelength and frequency

Sound is propagated as compressions and rarefactions (thinning) of the medium through which it is travelling, and is usually represented diagrammatically as a sine wave.

Wavelength is the measurement of the distance between the troughs (or crests) of two successive sound waves (Figure 45). The shorter the wavelength, the higher the frequency, i.e. the number of wavelengths per second. One sound wave per second is called 1 Hertz. Because sound travels in air at 340 m per second, 1 cycle per second (1 Hz) has a wavelength of 340 m. So, at 10 cycles per second (10 Hz), the wavelength is 34 m. In describing the frequency of bats' ultrasound, we refer to kilohertz, where 1 000 cycles per second is called 1 kilohertz (1 kHz). At a frequency of 50 kHz, which is average for bat echolocation calls, the wave is only 6.8 mm long and gives a distinguishable echo virtually the same size as the insect prey that the bat is hunting.

Human hearing ranges from 20 Hz to 20 kHz. While most bat echolocation calls are emitted at frequencies above the range of human hearing (i.e. ultrasound), we can hear the echolocation calls of some species. Echolocation calls of bats vary in frequency from 10 kHz to over 200 kHz. The calls of some free-tailed bats (e.g. *Otomops martiensseni*) and the Mauritian tomb bat (*Taphozous mauritianus*) are clearly audible to us.

Frequency modulation and bandwidth

A single note that does not vary in pitch is described as constant frequency (CF). A sound which varies in frequency is termed frequency modulated (FM). In bats with FM components to their calls, the frequencies typically sweep from high to low, but may also sweep from low to high. Since this change occurs in a few milliseconds (thousandths of a second), we do not hear the effect using heterodyne or frequency division detectors, although with time expansion detectors, it may be apparent. In practice, bats use a combination of FM and CF components in their calls.

The difference between the minimum and maximum frequencies in a call is called the bandwidth. By definition, a constant frequency call has a bandwidth of zero, while frequency-modulated calls have a non-zero bandwidth. Calls which span a small range of frequencies are termed narrow-band, while calls which vary greatly between their minimum and maximum frequencies are termed broad-band.

Harmonics

Most sounds, including the echolocation calls of bats, are complex mixtures of different frequencies. The different frequencies of complex sounds are not randomly ordered but follow a precise sequence of related frequencies. The lowest, or first harmonic, is usually termed the fundamental; further harmonics are exact multiples of the fundamental, i.e. x2, x3, x4, etc. Harmonics are usually weaker in intensity than the fundamental. Some bats, such as horseshoe bats, are capable of emphasising the second harmonic so as to increase the frequency of their call.

Sonograms

A sonogram is a graphical representation of sound, by means of a time-frequency graph, where the horizontal axis shows the duration of the sound and the vertical axis the frequency range. Sonograms are usually obtained by analysis of sound recordings made with a bat detector, using relevant computer software.

Intensity (i.e. amplitude or signal strength)

Intensity is expressed in units called decibels (dB), and for comparison between bats is usually measured at a fixed distance from a bat's mouth (10 cm). Intense echolocation calls measure 110 dB or more (equivalent in strength to a smoke detector alarm). Faint echolocation calls measure as little as 60 dB (the intensity level of a normal human conversation). Yellow house bats and little free-tailed bats are examples of high intensity, or 'shouting' echolocators, while common slit-faced bats are examples of low intensity, or 'whispering' bats. In many cases, call intensity is related to foraging behaviour. Bats that feed on the wing in open spaces produce the most intense calls. In contrast, those that feed by gleaning their prey off surfaces, relying on listening out for auditory cues, usually produce lower intensity calls.

Intensity can also vary within a single pulse; the yellow house bat uses a call which sweeps down from 50 kHz to around 30 kHz, but is most intense at its dominant frequency of around 38 kHz.

Call duration

Call duration varies from about 0.2 to 100 milliseconds (ms). The length of each echolocation call, and the rate at which it is produced (pulse repetition rate), varies from species to species, but can also vary within a species according to the situation. Call duration and pulse repetition rate vary considerably during foraging (Figure 46). Pulse repetition rate increases, and call duration decreases, as the bat locates and then attempts to catch its prey. During search phase, pulses may typically be emitted at around 10 per second. This increases to 40 pulses per second once the target has been located (approach phase), and may rise to 200 per second before the insect is caught (terminal phase, or 'feeding buzz').

Duty cycle

The echolocation calls of bats can be classified as high duty cycle or low duty cycle. In high duty cycle echolocation, individual calls are of longer duration than the inter-call intervals, and the echolocation signal is effectively switched 'on' most of the time the bat is echolocating. Such bats are capable of separating the emitted echolocation call and its echo in frequency, meaning that they can transmit and receive simultaneously. Separation of the call and its echo in frequency in high duty cycle bats is accomplished by means of an 'acoustic fovea' in the ear of the bat which is particularly sensitive to the precise frequency of the returning echo. Horseshoe bats and leaf-nosed bats tend to have high duty cycle calls. High duty cycle bats tend to have calls dominated by a CF component, although FM components are usually present too.

In low duty cycle calls, the emitted calls and their echoes are separated in time, so that duration of calls is much shorter than the inter-call intervals in order to give bats time to process the echo from one call before emitting the next call. Most bats, including plain-faced, sheath-tailed bats, free-tailed and slit-faced bats, have low duty cycle calls. Such bats usually have FM-dominated calls.

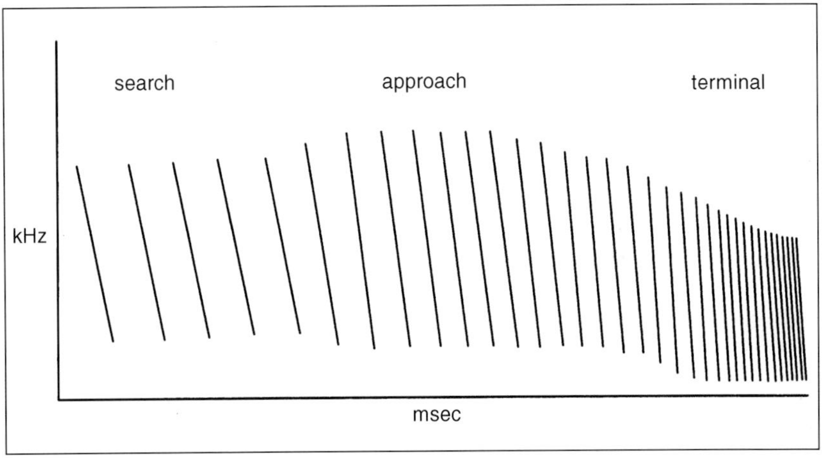

Figure 46 Diagram of a sonogram (time-frequency graph) depicting a bat's 'feeding buzz'

Doppler shift

Calls dominated by a CF (constant frequency) portion are usually of long duration, high pulse repetition rate, and high frequency, and often followed or preceded or both by a FM sweep. Such calls are capable of Doppler shift compensation, and are typical of families such as the horseshoe and leaf-nosed bats. The Doppler shift is best understood as the apparent drop in frequency that an observer hears when a fast car passes by. Doppler shift compensation in bats allows sensitivity to the flutter produced by the wing beat of moths and other insects.

Variability in call design

Since bats need to vary their foraging styles in different habitats, they must also change their echolocation calls. When identifying bats in the field using bat detectors, it is critical to bear in mind the habitat in which the bats are flying, as a species' echolocation calls may vary considerably between open and closed (i.e. cluttered) habitats.

Generally speaking, bats flying in open habitats have the following call features:

- lower pulse repetition rates;
- more intense, longer duration calls;
- narrower bandwidth;
- lower frequencies.

Bats of the same species flying in clutter would be expected to have the following call features:

- faster pulse repetition rates;
- less intense, shorter duration calls;
- broader bandwidth;
- higher frequencies.

EXAMPLES OF SOUTHERN AFRICAN BATS

To illustrate some of the parameters and concepts mentioned above, sonograms showing sequences of search phase echolocation calls are shown for species from three distinct families (Figure 47). These were all obtained by the author using the Pettersson D980 detector in conjunction with a Sony digital tape recorder and analysed by means of the Batsound programme running on a Pentium PC under Windows 95.

The sonogram of the bushveld horseshoe bat illustrates what is meant by a high duty cycle call. Call duration is 20 ms compared to 10 ms for the inter-call intervals, and the signal is thus switched on for 67% of the time. The bushveld horseshoe bat has a CF component to the call at precisely 82 kHz.

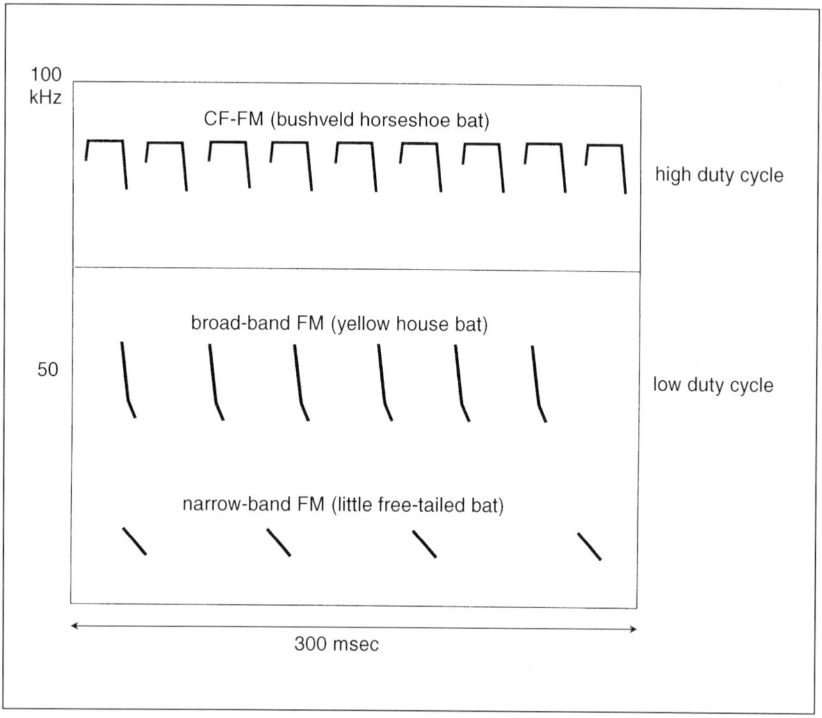

Figure 47 Diagram showing sonograms of southern African bats which use high duty cycle and low duty echolocation calls

Like most, if not all plain-faced bats, the yellow house bat has a low duty cycle call, with short call duration (about 6 ms in the example given), relatively much shorter than the inter-call interval. The echolocation signal is switched on for 12% of the time. Calls are emitted at a pulse rate of about 20 per second. The broad-band call is typically frequency modulated (FM), sweeping down from 60 kHz to 35 kHz. When viewed 'close-up', individual calls have a backward J-shape, and most of the energy is concentrated at the point where the steep part of the curve becomes shallow, i.e. around 38 kHz. With a heterodyne detector, the calls will sound like clicks.

The sonogram of the little free-tail bat (Figure 47) reveals the fact that free-tailed bats, like plain-faced bats, typically possess low duty cycle calls, which are only switched on for about 16% of the time the bat is echolocating. However, in comparison to the yellow house bat, the little free-tailed bat has longer duration (12 ms) calls which have greater tonal quality, sounding to a heterodyne detector more like a wet slap than a dry click. Frequencies are generally lower than for plain-faced bats, and the FM sweep is 'shallower', i.e. of lower bandwidth, from 23–28 kHz. The pulse rate in the example given is only 13 per second.

INTRODUCING BAT DETECTORS

The transformation of ultrasound into a signal audible to humans requires the use of specialised bat detectors, and can be achieved by one of three different methods which gives rise to three different categories of bat detectors: heterodyne, frequency division and time expansion. The type of detector system you purchase depends on what you want it for: fun, educational bat walks for the public, or for detailed acoustic field surveys. To assist readers in choosing and buying a detector, examples of different makes of bat detectors, showing the method of sound transformation as well as suppliers' details, are given in the Table on p. 168.

Heterodyne

With a heterodyne detector, the bat's ultrasound is converted into an electronic signal and compared with a variable internal signal, so that we listen to the (audible) difference between these two signals. Heterodyne detectors are 'tunable', allowing us to estimate the bat's actual frequency within a limited band of frequencies at a time, usually about 10 kHz. Disadvantages are that you have to manually scan through the full range of possible frequencies (although some expensive models are able to automatically scan frequencies for echolocation calls), and may miss bat passes in this way. Also, frequency information is lost if recordings are made.

In Europe, where the calls of local species are well known, heterodyne detectors such as the Magenta, Maplin, Batbox III and Mini-3 are very popular for educational bat walks involving the public, basic field identification of species based on audio output, and the monitoring of bat activity. Different models of heterodyne detectors have different tones. Once you have chosen a particular model, get to know it and stick to it. Different detectors also have different bandwidths. For example, if the bandwidth is ±10 kHz, and your bat detector is set at 30 kHz, you will hear frequencies from 25–35 kHz. Using headphones is a good idea, and will help you to concentrate better and to block out the noise from other bat detectors, as well as to get a better audible output over a wider bandwidth.

Frequency division

Frequency division detectors transform the entire spectrum of sound by producing one pulse per predetermined number of incoming pulses, say 8 or 10, thus lowering the frequency into the audible range. From their audio output alone (i.e. without recording equipment and analytical software), frequency division detectors are only capable of telling you that a bat is present, not at what frequency it is echolocating. However, they retain amplitude and frequency information, but not harmonic structure, in recordings, and are thus useful for recording and subsequent analysis of the sounds by computer, e.g. using the ANABAT system. The ANABAT system has been extensively used for acoustic species surveys by scientists, particularly in North America and Australia, and is competitively priced.

Time expansion with digital memory

Detectors capable of time expansion tend to be more expensive. Time expansion digitally provides the equivalent to recording sound on a high-speed tape recorder, and then playing it back at a lower speed. Time is effectively 'expanded' (slowed down) by a factor of, say, 10 (this factor can be changed in some models). For a 10-fold time expansion, a 3 second recording is replayed over 30 seconds. The advantage of time expansion is that all information, including amplitude, frequency and harmonic structure, is preserved, making this method ideal for subsequent, detailed analysis of call properties from recordings. Time expansion detectors are frequently used by scientists for acoustic species surveys, particularly in Europe. The sonograms shown in Figure 47 were obtained with a time-expansion detector.

INTERPRETATION OF HETERODYNE DETECTORS

When you listen to the sounds of bat ultrasound through a heterodyne detector, you will immediately notice a variety of different tones and rhythms used by different species. In Britain, bat-workers use words like 'thigh-slapping', 'stubble-burning', 'tap-dancing', and 'hair-rubbing'. On a recent local bat walk, a DBIG member coined the word 'marbles' to describe the 'feel' of the echolocation calls of Cape serotine bats.

What do these sounds actually mean? Understanding what these sounds mean, and being able to actually 'picture' the sonogram as we listen to the detector (much the same way we 'read' sonograms of the audible sounds made by frogs and birds) is the crux of developing the necessary basic listening skills. The following points are critical to bear in mind:

1) All bat identification begins with learning to differentiate between two types of calls: 'dry'-sounding clicks, and 'wet'-sounding smacks. Very short sounds resemble clicks on the detector, while progressively longer sounds acquire more tonal quality (smacks).
2) Remember that your heterodyne detector is very selective and has a narrow bandwidth of only 10 kHz, so that you are likely to be hearing only a portion of a bat's call at any one time. Of course, you can tune the frequency dial to hear different 'slices' of the call.
3) A single call may be composed of both faster (steeper) and slower (shallower) elements, so we may be able to hear both 'dry' and 'wet' components on a single call, and hence predict the shape of the call (i.e. picture the shape of the sonogram). For example, bats like Schreibers's long-fingered bat, yellow house bat, Cape serotine, and banana bat have calls comprising a steep-FM portion ('dry' clicks) followed by a shallower tail portion ('wet' smacks).
4) High duty cycle echolocation calls with a dominant constant frequency (CF) component are the easiest to identify, as you will be able to tune in precisely to the constant frequency, and the sound will have a distinctive high, almost musical quality.

Steps to identification

Start your search with the detector on medium volume and set at 45 kHz. Most bat calls with FM components sweep through this frequency.

Try to picture the sonogram by imagining tuning your detector up and down the curve. Is it a vertical sweep (dry clicks throughout the frequency range)? Does it flatten out at the end? Is it a single note throughout (as in CF bats like horseshoe bats)?

Be aware of the habitat and behaviour of the bats. Calls are most species-specific in the search phase, so feeding bats, or bats just emerging from their roost, although easier to hear, may be misleading. Step back some 20–30 m from feeding areas to listen to their calls as they approach or leave a feeding area.

Ask yourself the following questions when making an identification:

1) What is the best listening frequency?
2) Are you listening to 'dry' clicks or 'wet' smacks?
3) What is the rhythm of the call?
4) How loud is the bat?
5) Is there a point where the clicks turn to smacks?
6) At what frequency does this occur?
7) How fast are the pulses coming?
8) What is the frequency range?
9) Is there the same amount of energy throughout the whole of the call?

Determining the zero point

As mentioned above, many low duty cycle bats have a characteristically-shaped call with a steep portion which flattens out at the bottom, much like the letter 'J' backwards. The bottom of the 'J' represents the end point. This point was useful

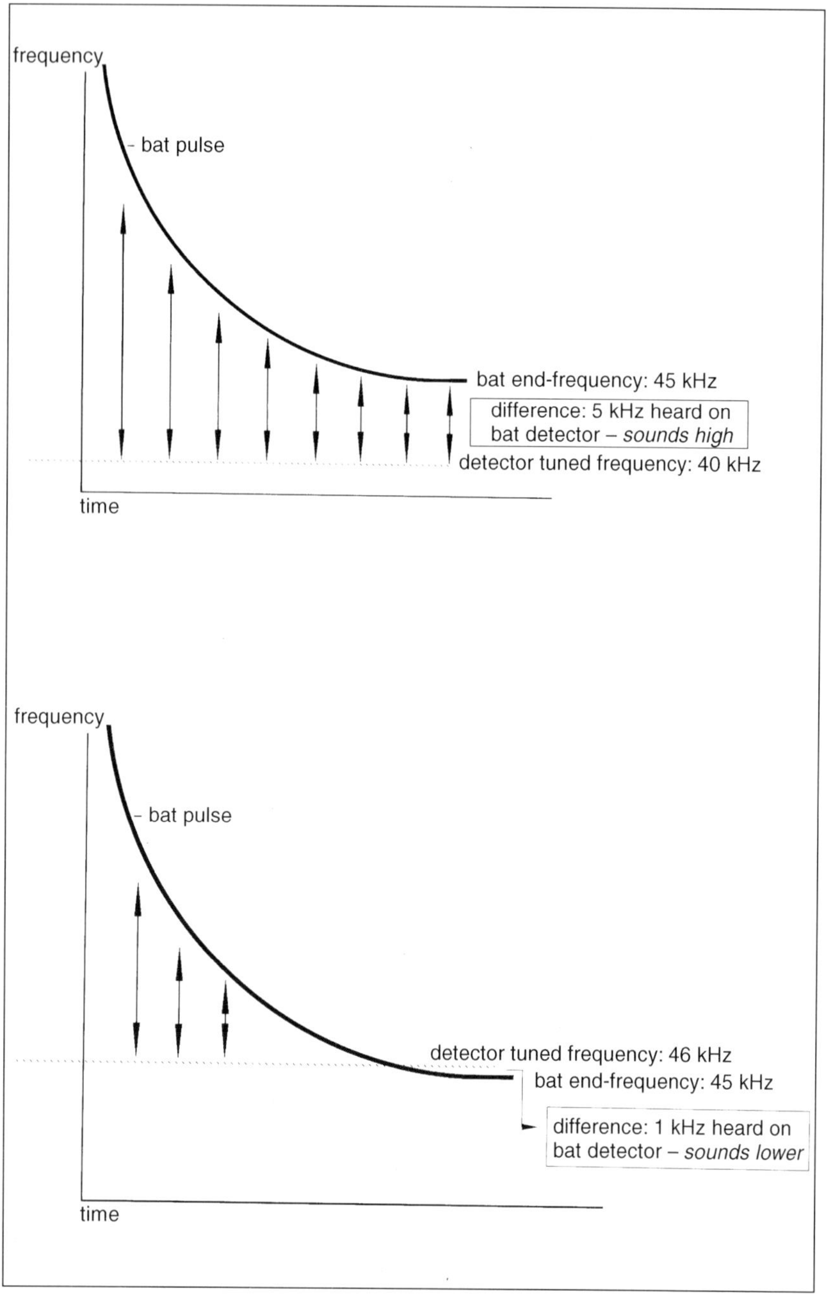

Figure 48 Diagram explaining the principle of zero point

for differentiating between two sonic types of the common European pipistrelle, with end points of 45 kHz and 55 kHz. Many of our common southern African bats have similarly-shaped calls, for example, Schreibers's long-fingered bat, Cape serotine, banana bat, and yellow house bat. The procedure in Figure 48 will prove most useful for determining their end points accurately for identification purposes.

From Figure 48 we can see how to use a heterodyne detector to determine the position of the 'tail' using the concept of the zero point. Recall that a heterodyne detector measures the difference between an internal signal (frequency to which the dial is set) and the bat's signal. We change the internal frequency when we tune the dial. The resulting audible 'click' we hear represents the difference between the two frequencies. If our detector is tuned to 40 kHz and the bat is echolocating at 45 kHz, we listen to the difference between these two frequencies, which is a sound at 5 kHz, well within our range of hearing (remember, humans hear from 20 Hz to 20 kHz). When our detector is tuned exactly to the bat's frequency of 45 kHz, there should theoretically be no sound, i.e. the difference is 0 kHz. In reality, because bat sounds are composed of a range of harmonic frequencies, we always hear something.

But it is possible in practice to find this zero point. If the bat's end point is 45 kHz, and our detector is tuned to 40 kHz, we are listening to a sound of 5 kHz. As we tune the detector upwards, the difference between the tuned frequency and the bat's frequency becomes smaller, so that when our detector is tuned to 46 kHz the difference is only 1 kHz. Continuing upwards, when our tuned frequency reaches 50 kHz, the difference between this and the bat's end point frequency of 45 kHz is once more 5 kHz, i.e. it is pitched higher than when the difference was 1 kHz. By dialing the tuned frequency up and down until the 'wet' smack sounds lowest, we can get close to the zero point.

The following table provides details of suppliers of a range of bat detectors. Further details and online orders of bat detectors can be found at the websites of Bat Conservation International and the Bat Conservation Trust (see Important contacts). The models listed in the table following do not necessarily imply the author's endorsement of any of these models, nor his lack of endorsement of any models inadvertently omitted.

Detector make	Technique for transformation	Supplier's details
Magenta DIY kit (or built-up unit)	Heterodyne	Magenta Electronics Ltd 135 Hunter Street Burton-on-Trent, Staffs DE14 2ST, UK Tel: 0944 1283 565435 Fax: 0944 1283 546932 **www.magenta2000.co.uk**
Batbox III	Heterodyne	Stag Electronics 4 Esprit Court, New Road Shoreham-by-Sea West Sussex, BN43 6RB, UK Tel/Fax: 0944 07000 228269 e-mail: **info@batbox.com** **www.batbox.com**
Mini-3	Heterodyne	Ultrasound Advice, London Can also be ordered through Bat Conservation International (see full details on p. 186) **www.batcon.org**
Maplin kit	Heterodyne	Hertfordshire and Middlesex Bat Group (Dinah Harding) 2 Stonelea Road, Hemel Hempstead, Hertfordshire HP3 9JY, UK Tel: 0944 1442 252195
ANABAT	Frequency division	Titley Electronics P.O. Box 19 Ballina NSW 2478 Australia Tel/Fax: 0966 86 6617 e-mail: **corben@delphi.com**
Pettersson D980	Heterodyne Frequency division Time expansion	Pettersson Electronik AB Tallbacksvagen 51 S-756 45 Uppsala Sweden Tel: 0946 1830 3880 Fax: 0946 1830 3840 e-mail: **pettersson@bahnhof.se** **www.bahnhof.se/~pettersson**
Tranquility	Time expansion Heterodyne	David J. Bale 3 Suffolk Street Cheltenham Gloucestershire GL50 2DH, UK Tel/Fax: 0944 1242 570123 **www.users.globalnet.co.uk/ ~courtpan**

Chapter Nine

Bats in the belfry

For a variety of reasons, many householders are keen to evict their bat tenants. Concerns relating to nuisance factors such as the smell and noise associated with large colonies of free-tailed bats are sometimes genuine. However, the majority of householders complaining of 'problem bats' harbour irrational and unfounded fears about perceived disease risks, or the fear of aggressive attacks by bats (see Chapter Two). All callers who telephone the Durban Bat Interest Group (DBIG) to complain about 'problem bats' are sent promotional information on bats, and some receive a personal visit from trained bat-workers. Many callers, reassured of the innocuous and even beneficial nature of bats, are persuaded to tolerate, if not welcome, their bats. Those who refuse to tolerate their bats, or who have genuine nuisance problems that cannot be solved practically, are advised against the use of lethal fumigation, and provided with advice on bat-proofing (see information following), as well as referred to reputable pest control companies capable of conducting non-lethal bat-proofing.

The aim of this chapter is to provide readers with responsible advice on problem roof-bats, based on trends and research undertaken worldwide. Since this is such a detailed and technical topic, it is only possible to provide general guidelines and principles. The reader is referred to excellent technical manuals (see Further technical advice, p. 176), or contacts where they may be referred to reputable pest control companies for further advice (see Important contacts).

Before reading this chapter, re-read Chapter 2 to make sure that your fears are not based on irrational myths! The best choice is almost always to keep your bats. If you have to exclude, read on. Consider putting up bat houses prior to excluding the bats to accommodate the evicted tenants (see Chapter Ten).

PROTECTION OF BATS

In order to protect bats from unscrupulous extermination, we need to understand how existing laws can be used to enforce protection of bats; also how they can be improved to better protect bats. The information following is based on an understanding of the South African situation, which is assumed to be similar in other southern African countries. Readers from other countries should contact their local national parks department, or other conservation NGOs for details on laws regarding protection of bats and the use of pesticides in bat control.

While bats are afforded legal protection in countries like Britain (Wildlife and Countryside Act of 1981–1982), other European countries (11 European nations signed an Agreement on the Conservation of Bats in Europe in January 1994), and the United States (where it is illegal in most states to use poisons to kill bats), no such legal protection exists for bats as a whole in South Africa, or, to the best of my knowledge, other countries within southern Africa.

However, no pesticide in South Africa is registered for use against bats, which means that it is illegal for pest control operators to fumigate bats using methyl bromide. This important fact should be used to full effect by South African bat-workers seeking to prevent unnecessary exterminations of roof-bat colonies.

In practice, fumigation of bats by pest control companies still continues, fuelled by an ignorant public and a minority of profit-motivated, unscrupulous pest control operators.

THE *OTOMOPS* STORY

In October 1996, two species of bats, the large-eared free-tailed bat (*Otomops martiensseni*) and the short-eared trident bat (*Cloeotis percivali*), were proclaimed by the KwaZulu-Natal government as Endangered Mammals (Schedule 6 of the 1974 Provincial Ordinance), affording them legal protection, in the form of fines of up to R500, or a jail sentence, for persons caught killing or otherwise 'disposing of' one of these bats without a permit. In 1999, a new provincial act ushered in a new Conservation Ordinance for KwaZulu-Natal, and it is encouraging that the two species listed above are listed under the highest priority Specially Protected category, along with species such as the black rhino. A number of other bat species are listed under the Protected category, where they will be afforded legal protection against human disturbance.

Within South Africa, the large-eared free-tailed bat occurs only in the roofs of houses in the Durban area where its long-term survival may be threatened by ongoing fumigation of houses against wood borer. Following the provincial government legislation mentioned above, and resulting from press releases and letters written to pest control companies in the Durban area in late 1996 and early 1997, pest control companies now should report all bat-related complaints to the Durban Natural Science Museum or the KwaZulu-Natal Nature Conservation Service, for referral to companies qualified to perform humane bat evictions, and instances where householders fumigate their houses specifically to eradicate bats seem to be in sharp decline in the Durban area.

The KwaZulu-Natal situation provides a model whereby provincial or national government legislation can be a crucial lever in changing detrimental pest control or other human disturbance practices which may be seriously threatening the survival of certain species.

Wider countrywide efforts in South Africa, and other southern African countries, aimed at the public and pest control industry, are required to promote greater legal protection and awareness about bats, and the correct non-lethal exclusion procedures. It is imperative that bat-workers liaise closely with the pest control industry in a non-confrontational manner, to negotiate a better deal for bats. This approach worked extremely well in the United States, to the extent that one pest control association devoted an entire issue of its magazine to articles on humane bat eviction, written by bat conservationists. It has also been successful in Britain, where the bat lobby has encouraged the pest control industry to move away from lethal timber treatment pesticides such as lindane, in favour of pyrethroid-based products which are harmless to bats.

Locally, this approach is beginning to work in the Durban metropolitan region, where 'problem bat' colonies are now being reported to DBIG when located by pest control companies prior to roof fumigations for wood borer. Unfortunately it is difficult to measure how many cases are not reported, and where fumigations or irresponsible evictions (e.g. when young are present) may still be occurring. At least two roof colonies of the endangered large-eared free-tailed bat have been saved from certain fumigation by the intervention of alert and concerned pest control company personnel who recognised the bats prior to methyl bromide fumigation, and notified DBIG, who then removed the bats.

FUMIGATION BY METHYL BROMIDE

In South Africa, the highly toxic gaseous fumigant, methyl bromide, is widely used in conjunction with a tent to eradicate insect pests, and in particular wood borer. Although it is not registered for the control of bats, and its use for bat extermination is strictly illegal, methyl bromide has been used to exterminate bats in the past, and this probably continues illegally in the present.

According to Dr Merlin D. Tuttle of Bat Conservation International, in a letter from him addressed to the Durban Bat Interest Group:

> Methyl bromide is highly dangerous to people as used in bat fumigation and has little effect in solving nuisance problems. It can seriously damage insulation on electrical wires, creating potential fire hazards, and collects in the lowest parts of a building, where it is more likely to poison people than bats. If roost entries are closed to exclude bats, fumigation is pointless, and if exclusion is not practised, the site will be promptly re-colonized. For these reasons, this and all other toxicant use against bats is now illegal in the US . . . All bat poisoning can be considered fraudulent and counterproductive.

While methly bromide is banned in some countries, and its banning has been considered in South Africa, a legal obstacle to banning methyl bromide use is the fact that three species of local wood borer are notifiable pests, meaning that one is legally obliged to take action whenever a borer is located in a residential dwelling, whether or not bats are present. Methyl bromide is apparently the only registered chemical which is effective in the eradication of borers. However, alternative chemicals with low mammalian toxicity should be sought as is the case in Britain, and conservationists should lobby for stricter punishment for offenders who continue to use methyl bromide illegally to fumigate bats.

TIMBER TREATMENT

All timber in modern houses (after the 1950s) is pre-treated. Older houses with untreated timbers are most commonly borer-infested, but, ironically, these are the same houses that are structurally best suited to bats.

Insect infestations can be controlled through the direct application of pesticide fluids to timber. One of the most common ingredients of pesticide fluids applied

to timber, an organochloride called lindane (gamma HCH), has been shown to be highly toxic to bats (and people). In experimental cages treated with lindane some 14 months previously, British researchers found that bats died in a matter of days due to acute poisoning. Similar mortality was observed when the cages were treated with a common fungicide, pentachlorophenol (PCP). Other organochlorides used in the pesticide industry have a sub-lethal, chronic effect on bats, accumulating in their bodies over time, and ultimately killing them. In Britain, a much safer synthetic pyrethroid-based compound called permethrin is now widely used, which has a low toxicity in mammals, and the use of permethrin and other such 'bat-friendly' pesticides should be actively promoted.
Warning: As a rule, bat-workers should allow a minimum period of at least six months following roof treatment with lindane before entering a roof.

BAT GUANO

Guano accumulations from large bat colonies in roofs can create serious problems. Apart from the obvious odour problem, large accumulations of guano can cause ceiling boards to buckle under the weight. Rainwater leaks can also result in the guano forming solid, wet layers which can cause rotting of older makes of ceiling boards. Since guano usually accumulates in a restricted area of the roof where the bat colony is located, householders can be advised to remove the guano, using an industrial vacuum cleaner, and then carefully place a tarpaulin or tray of sorts to collect further accumulations of guano so as to make it easier to remove on a regular basis.

On the plus side, home owners with guano problems can be reminded that bat guano makes excellent garden fertiliser. I have a 210 litre drum filled with liquidised bat guano, and invite anyone to come and see its effects on my herbs, cycads, indoors plants and bonsai. In many countries guano is mined and marketed as fertiliser for a high price. One local pest control operator reported removing 17 black garbage bags full of guano from a single roof. Local entrepreneurs could obtain substantial quantities by offering to remove the guano from grateful householders, and attract the interest of the larger nurseries to market the bat guano, thereby placing a real monetary value on having lots of bats in your roof.
Warning: In view of the remote chance of contracting histoplasmosis, it is recommended that anyone cleaning up bat guano in roofs, or visiting caves or roofs with large accumulations of bat guano, wear the appropriate pathogen filter mask (with pore diameter of 1 micron), and avoid inhaling dust that may contain the fungal spores which can become airborne when the guano is disturbed.

REPELLENTS

In his book, *America's neighbourhood bats,* Dr Merlin D. Tuttle writes 'Unfortunately, no ultrasonic or chemical repellents appear to be moderately effective in eliminating bats from their day roosts in buildings.' Putting lights in the roof or increasing the ventilation are never more than partially effective. Burning sulphur candles or smoke bombs in roofs may cause the bats to temporarily vacate the roof, but then the exit holes should be sealed to prevent the bats returning. With sulphur candles, adequate fire hazard precautions should be taken. Dog and cat spray aerosols can be used to deter bats from particular night roosting spots for up to several months. For humane reasons, the roosting surface and not the bats should be sprayed.

BAT EVICTION

In 1935 in the United States, J. Silver wrote:

> The most satisfactory and the only permanent way of obtaining freedom from the batroost nuisance . . . is to shut the bats out. This may be called 'batproofing' the building.

Over 60 years later, this is still the best advice. The only effective way to permanently get rid of bats in roof situations is to exclude them by blocking up all the exit holes when the bats are absent. Remember, evictions should never be done when flightless young may be present (i.e. September to May in South Africa).

Some of the basic steps involved in confirming the presence of bats in a roof, and undertaking the eviction, are given. However, every house and situation is different, and evictions may prove to be technically complicated, e.g. a roof space with a corrugated iron roof where the corrugations provide a seemingly unlimited number of possible exit holes, or a thatched roof where it is almost impossible to seal every crack between the thatch and wooden poles. Such cases should be referred to a professional (see Further technical advice on p. 176).

Finding roof bat colonies

The first step is to find out whether bats are actually present. Quite often, a householder will swear to having bats in the roof because of nocturnal scratching noises, which turn out to be caused by rats or even birds. I was recently notified of bats 'infesting' a toilet in a factory. 'They've even laid their eggs,' I was informed! The culprits turned out to be Indian mynas. Or they may have observed bats flying around in their yard and assumed these came from the roof. In other cases, a previously reported bat colony may have moved away or even been inadvertently poisoned. To confirm whether bats are present in a roof, follow a simple three step procedure:

1) Ask the occupier if they have seen bats emerging from the roof, or have any other proof of bats;
2) Check for the presence of bats by listening for their chittering noises in the roof. Shining a torch on areas where bats are likely to be hiding may help to elicit a response (remember that you will never see some of the smaller bats during the day);
3) Always look out for tell-tale piles of scattered bat guano (droppings). Bat droppings are irregularly-shaped and not pointed at each end like a rodent dropping. They do not have the white uric acid tips characteristics of gecko or bird droppings. They also crumble easily into fine chitinous insect parts.

As a rule, bat colonies in roofs are localised in one area of the roof, although bats roosting between the tiles and roof lining may sometimes move around over a wider area of the roof. The first and favourite place to look for a bat colony is on the gable apex wall, where bats are often tucked up between the brick wall and rafters. Piles of droppings at the base of the gable apex are usually the first sign of such a colony. Other common roost sites include above the eaves, on top of the ridge beam and other rafters, underneath the ridge tile, between tiles, and in broken air vents.

Finding and closing exit holes

Exit holes can usually be located by watching at dusk to see where bats emerge. Closer inspection by day will reveal the exact nature of the openings. In most cases, bats seem to roost close to where they emerge and enter the roof, but some colonies will crawl some distance between their preferred roost site and the point of emergence.

Common places to look for access points will usually be under the eaves where the roof meets the wall, at the gable apices, beneath the ridge tile, between tiles or corrugated iron sheets, or even between the slats of a broken air vent (see Figure 49). According to one bat-proofing expert, the smaller species, like little free-tailed bats, can gain access through the tiniest opening: the 'rule of thumb' is that if you can fit your thumb through the opening, then it is large enough for a bat.

It is also possible to detect exit holes from within the roof, by detecting faint air currents caused by even the smallest openings. One ingenious device for detecting these air currents involves a sheet of thin plastic (Gladwrap), or even a tissue, joined by clothes pegs or tape to the two ends of a clothes hanger.

Removing the bats

Exclusion can be done at dusk after the bats have emerged to feed. Most of the openings can be sealed with a foam-spray masonry sealant. However, one or two holes should be closed temporarily (using chicken wire, newspaper, or some similar material) and reopened before dusk on the following two days to allow any bats that remained behind on the first day to get out. Do not forget to close holes again on each of the following two days before dawn. On the third day, or when no more bats are seen leaving at dusk, the final hole can be permanently sealed. Another effective option is to use polypropylene bird netting or plastic sheeting (Figure 50). Hang the netting during the day directly above remaining unplugged exit holes, extending at least 30 cm to each side and below, using duct tape or staples. The sides may be attached but the bottom must hang free. This allows the bats to drop down to leave, but when they attempt to fly straight in upon their return, the netting acts as a one-way exclusion valve. Allow at least two or three warm days for all the bats to leave and then plug the exit hole securely. Some people have suggested smoking out all the bats at the same time, and then plugging the holes; however, this approach is not recommended as great care should be taken to ensure that the smoking device, or 'fogger', does not contain a lethal fumigant which would ultimately kill the bats.

There are other ways of removing bats from the roof space. Some, such as large-eared free-tailed bats and yellow house bats, can be caught by hand in the roof, but this should not be attempted without gloves. Various traps have been designed to catch bats as they emerge. One such trap is the 'harp trap' (Figure 51), originally designed by D.G. Constantine and modified by M.D. Tuttle. An ingenious homemade device featured in the *Old House Journal* of 1980, uses a material funnel covering the exit hole, connected by a metal tube too slippery for bats to climb up, and then connected in turn to a bag with slippery sides to collect all the bats as they drop out of their roost at dusk.

The process of bat eviction described above can often be technically difficult, for example where numerous exits are present. In such cases, a responsible pest controller should be approached to bat-proof the premises. One of the existing bat interest groups can be contacted for details of suitable pest control companies (see Important contacts).

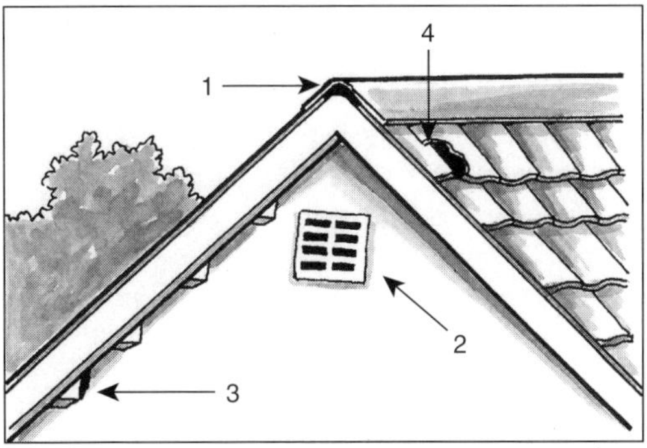

1. apex of roof (under ridge cap)
2. air vents
3. gaps near rafters
4. broken tile

Figure 49 Common access points for bats occupying roof spaces of buildings

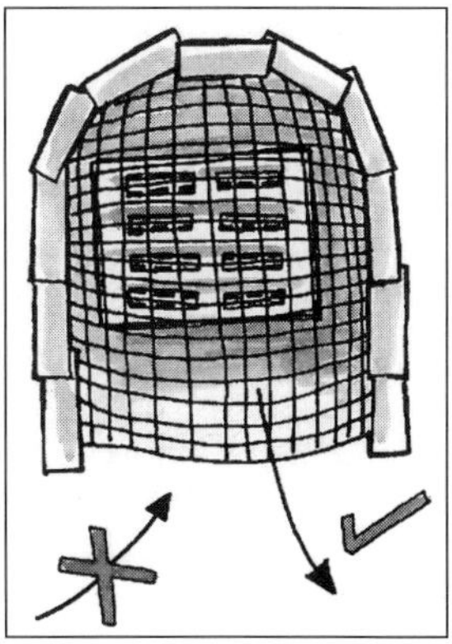

Figure 50 One-way valve used to exclude bats from buildings

Figure 51 Harp bat trap

Providing alternative accommodation

Whenever doing a bat exclusion, home owners should be strongly encouraged to erect bat houses to accommodate the evicted bats. This will prevent them from simply moving next door and afflicting your neighbours with the problem. Bat houses should be erected some time before the bat exclusion so as to familiarise the bats with the bat house. Refer to Chapter Ten for more detail on bat houses.

FURTHER TECHNICAL ADVICE

Detailed procedures for bat-proofing properties are given in the publications, Hutson, *Bats in houses* (1986, British Fauna and Flora Preservation Society), *House bat management* (Resource Publication No. 143, United States Fish and Wildlife Service), and Williams-Whilmer and Brittingham, *A homeowner's guide to Northeastern bats and bat problems* (1995, Pennsylvania State University). A 23 minute video entitled 'Bat-free belfries: A guide to bat-proofing' is available from the last-mentioned publisher. Publishers' addresses for the above publications are listed in the Bibliography. As yet, no handbook is generally available for South Africa.

Chapter Ten

Making bats welcome

WHY BAT HOUSES?

Just what are bat houses, how successful are they, and how should they be constructed to attract southern African bats? These are questions more and more people are asking, particularly in the agricultural sector where real economic returns are anticipated with diminished crop losses to insects. Bat houses are also attracting much interest as ecotourism features associated with game lodges and nature reserves.

With a growing emphasis on biological control and integrated pest management, more and more farmers, and the public, are taking bats seriously, especially since a single colony of bats may consume millions of insects, including crop pests, each growing season. Instead of eradicating bat colonies from their farmhouses, farmers are now exploring ways of attracting bat colonies to their fields by erecting bat houses.

Bat houses should be used as a matter of course whenever bat-proofing of buildings take place, so as to accommodate the evicted bats.

Research on bat houses in North America has shown that their use can significantly aid in rebuilding healthy populations of valuable bat species. The North American Bat House Research Project involves some 5 000 volunteers who experiment with bat houses. Small design, location and temperature differences can substantially increase success. Much useful information on bat houses and the North American Bat House Research Project can be found at Bat Conservation International's (BCI's) website **http://www.batcon.org**. The latest edition (1997) of *The Bat House Builder's Handbook*, by M.D. Tuttle and D.L. Hensley, is well worth obtaining.

There is every reason to believe that the widespread use of bat houses in southern Africa will greatly aid in the conservation of our bat fauna, as well as benefit farmers through the pest control services of resident bat colonies. There is also every reason to believe that North American-designed bat houses should work well in southern Africa, although this remains to be tested by large-scale experimentation. One of the groups of bats most readily attracted to bat houses in many parts of the world are the crevice-dwelling free-tailed bats (family Molossidae). Small to medium-sized free-tailed bats occurring in both North America and southern Africa have broadly similar climatic and roosting preferences.

Environmental requirements appear to be more important than bat house design *per se*, so it is absolutely essential to test a few simple designs under a range of environmental conditions. Only with large-scale experimentation involving the wider public will we establish the criteria for successful occupation of bat houses by our different groups of local bats.

HISTORY OF BAT HOUSES

The first bat houses were giant 'towers' built in Texas in 1900 by Dr Charles A.R. Campbell, as reported in his book *Bats, Mosquitoes and Dollars* (1925). These towers were built to attract large Mexican free-tailed bat colonies to control malaria-bearing mosquitoes. Although the success of this grand experiment cannot be proven, the substantial guano deposits from the colonies were commercially harvested as prized fertiliser. Bat houses have been widely used in North America for at least the last 30 years.

The first description of a European 'bat box' was published in France in 1918. Within 20 years, a number of experiments were carried out in Europe with introductions of a large number of bat boxes in conifer plantations. In Britain, the first bat boxes were positioned in 1968. In a later project, sponsored by BBC TV, the Forestry Commission, and the World Wildlife Fund, over 3 000 boxes were placed in six forests from northern Scotland to southern England, with very encouraging results.

In South Africa, far-sighted South African Railways authorities supervised the construction of two huge bat houses at Komatipoort in the mid 1950s, also in the hopes of controlling malaria. One of these is still occupied by a large colony of Angolan free-tailed bats today, while the second was recently deliberately poisoned and is unoccupied.

OBTAINING A BAT HOUSE

Nigel Fernsby of 'Birds and Bats – Environmental Products' and a member of the Gauteng Bat Interest Group, has made a promising start in South Africa with the construction and erection of locally manufactured bat houses. Five distinct designs are now available for purchase, suited to different species of southern African bats, ranging from the 'crevice-dwellers', such as free-tailed bats, to 'cubicle-dwellers', like serotine bats and yellow house bats.

Farmers and ecotourism resort owners from the Western Cape, Eastern Cape, Gauteng, Mpumalanga, KwaZulu-Natal, Mozambique and Kenya have already bought or shown an interest in Fernsby's bat houses. Although a period of up to a year can be expected before bats inhabit a suitable bat house, some of Fernsby's 'Angolan free-tailed' bat houses attracted occupants within a few weeks. A large 'bat hotel' erected by Fernsby at a timeshare resort on the edge of the Kruger National Park has proved successful in accommodating a large colony of bats which previously occupied the corrugated iron roof over a game viewing deck where their droppings and odour were regarded as a major nuisance by guests.

For those wishing to experiment with designing and constructing their own bat house, bat houses can be constructed very easily and with minimal cost. Two simple but successful (at least in North America) designs recommended by Bat Conservation International in their *Bat House Builder's Handbook*, are shown here, termed the 'Small economy' (Figure 52) and 'Nursery' (Figure 53) bat houses. Being larger in size, the 'Nursery' design may be expected to attract larger nursery colonies (breeding females and their young) of crevice-dwelling bats, most probably, and depending on your location, little, Angolan or Egyptian free-tailed bats.

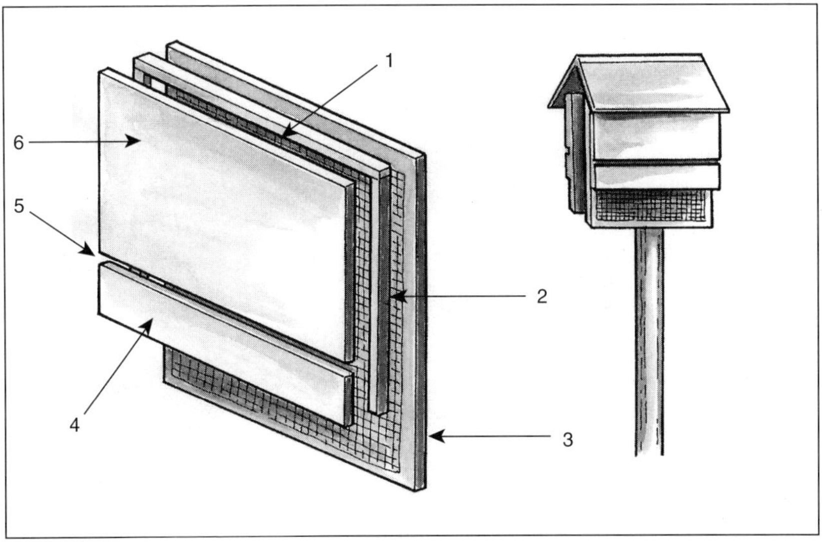

Figure 52 Construction of Bat Conservation International's Small Economy House (Tuttle 1988). This bat house design can be made from a 635 mm × 1 270 mm piece of exterior plywood (front, back, and roof) and one piece of 18 mm × 44 mm × 1 625 mm shutter plywood (spacers). A 63 mm × 18 mm × 610 mm board makes a roof. This bat house works especially well when mounted on buildings or in pairs back to back on a single pole, in which case a single roof can be built over both houses. It can hold 50 or more bats. (See caption for Figure 53 for important tips about mesh, caulking, and painting requirements.)

1. Top spacer 44 mm × 610 mm
2. Side spacers (×2) 44 mm × 508 mm
3. Back 673 mm × 610 mm
4. Lower front 127 mm × 610 mm
5. Vent 12 mm × 610 mm
6. Upper front 419 mm × 610 mm

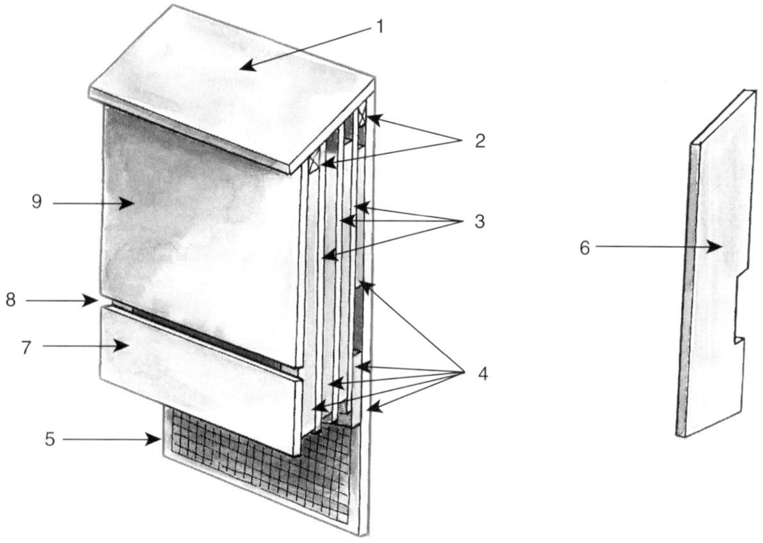

Figure 53 Construction of Bat Conservation International's Nursery House (Tuttle 1988). This house is designed to meet the needs of a variety of crevice-roosting bats and will house up to 200. The roof, front, and back pieces are made from 12 mm exterior plywood and the partitions from 6 mm plywood. The sides, spacers, and roof supports are 18 mm shutter plywood. Houses built with screws instead of nails last much longer. Staple heavy-duty plastic mesh to the front side of each partition and inside the back, extending across the landing platform. 3 mm mesh similar to Internet #XV-1670 (1-800-328-8456) is recommended. It is very important to caulk all exterior joints to prevent heat loss and moisture leakage. Paint your finished bat house with two or three coats of exterior latex. Darker colours may be necessary in cooler southern temperate climates in southern Africa, but medium colours are often better in warm, tropical and subtropical areas. Vents are not needed in the coolest locations. Buildings and poles are the preferred mounting locations.

1. Roof 508 mm × 165 mm
2. Roof supports 406 mm × 38 mm
3. Partitions 406 mm × 546 mm
4. Spacers (located between partitions along sides)
   (6) 508 mm × 18 mm (2) 254 mm × 18 mm (3) 127 mm × 18 mm
5. Back 444 mm × 775 mm
6. Sides 100 mm wide, 585 mm tall at front and 635 mm at back
7. Lower front
8. Vent 444 mm × 12 mm
9. Upper front 444 mm × 432 mm

SITING AND MOUNTING YOUR BAT HOUSE

As most bats need a drop, and since free-tailed bats particularly, tend to fly in open spaces above the vegetation canopy, bat houses should be placed at a height of at least 5 m. Smaller bat houses can easily be mounted on the wall of a building under the eaves, leaving a clear space for the bats to leave and return unhindered, or they can be mounted on a single pole. It is a good idea to mount two bat houses back-to-back facing north and south (Figure 52), with slots in the backs of each to allow bats to move between them and even roost in the space created between them. This system creates a temperature gradient between the warmer north-facing and cooler south-facing bat houses, thus allowing the bats a range of conditions to select from. The gradient can be exaggerated further by painting the north-facing bat house black to absorb more of the sun's energy, while the south-facing bat house is painted white. Larger bat houses, such as Fernsby's 'Angolan free-tailed' bat house, can be mounted across two sturdy poles.

Untreated timber should be used on the inner surfaces, and bat houses should be sealed to prevent draughts or water leaks. Bats do not appreciate leaky or draughty living quarters any more than humans do! Bat houses are commonly open-bottomed to allow access to bats, and not birds or other animals, although Fernsby's local designs provide openings under the 'eaves' of his bat houses, as well as bottom flaps to allow removal of guano accumulations.

Bat house owners should use a combination of different-sized bat houses to attract a diversity of different-sized bat species which will prey on a wider range of insect species. Your chances of success are greatly improved by putting up more than one bat house at a locality, preferably facing different directions to receive different amounts of sun, providing a range of temperatures and environmental conditions for bats. As already mentioned, no one position is optimal, and it is important to experiment with a range of conditions. Not only do we have both cool and hot climates in southern Africa, but different species have very different requirements. Bats that readily become torpid, such as yellow house bats, may prefer colder temperatures during winter, while at least some species of free-tailed bats generally prefer warmer temperatures throughout the year. Even within one species, temperature preferences may vary greatly throughout the year, between hibernating and maternity colonies for example.

ATTRACTING BATS TO YOUR GARDEN

A few species of fruit-eating megabats are commonly found in urban and suburban areas. In Durban, colonies are even found in heavily built-up shopping areas, where they may roost on tall alien trees (like royal palms) or man-made structures such as under the eaves of tall buildings. In Cape Town, cave-roosting Egyptian fruit bats are regarded to be a nuisance by residents of the southern suburbs, where they feed on yellow-wood trees and defaecate on walls and cars. Residents complaining of fruit bats defaecating on their walls should try illuminating the walls with a bright light, as this has been found to be successful.

You can attract fruit bats to your garden by planting either their favourite indigenous trees (see list following), or soft-fleshed fruit trees like bananas, guavas, pawpaws, and mangos. It is also possible to attract fruit bats to your bird table at night by putting out bananas and other fruits, but first make sure you have fruit bats in your suburb. The mournful and monotonous, honking, male territorial call of epauletted bats is one way of knowing you have fruit bats in the vicinity, but these species are only found in the eastern and northern parts of southern Africa.

Remember that planting indigenous trees, shrubs and ground covers, as well as constructing water features, will increase the supply of insects which in turn will attract more microbats. In Britain, nurseries actually supply a 'bat mix' of wildflower seeds which germinate into plants which are known to attract moths and other insects favoured by insectivorous bats.

Beware of spraying fruit trees with insecticides which could easily contaminate both fruit-eating megabats, and insectivorous microbats, not to mention birds.

Indigenous trees favoured by Wahlberg's epauletted fruit bat

The following list was provided by C. Sapsford, formerly of the University of Natal, who conducted a study of the diet of Wahlberg's epauletted fruit bats based on examination of 'spit-outs' collected from under their feeding roosts:

Cape fig (*Ficus sur*)
Swamp fig (*Ficus trichopoda*)
Natal fig (*Ficus natalensis*)
Wild frangipani (*Voaconga thouarsii*)
Forest toad tree (*Tabernaemontana ventricosa*)
Water pear (*Syzigium cordatum*)
Coastal goldenleaf (*Bridelia mucrantha*)
Natal ebony (*Euclea natalensis*)
Dune myrtle (*Eugenia capensis*)
Cape ash (*Ekebergia capensis*)
Wild custard-apple (*Annona senegalensis*)
Yellow-wood (*Podocarpus latifolius* and *falcatus*)
White milkwood (*Sideroxylon inerme*)
Quinine tree (*Rauvolfia caffra*)
Tree fuschia (*Halleria lucida*)
Marula (*Sclerocarya caffra*)
Natal mahogany (*Trichelia emetica*)
Wild plum (*Harpephyllum caffrum*)
Red milkwood (*Mimusops caffra*) Highly favoured

FURTHER TECHNICAL ADVICE

More details on bats, bat houses, and the North American Bat House Project can be found on the Internet at Bat Conservation International's (BCI) website: **http://www.batcon.org/**

The excellent *Bat House Builder's Manual* is also available from BCI. Mr Fernsby's contact details are given under Important contacts. Bats houses can be ordered from Mr Fernsby for prices (at time of going to press) ranging from R160 for the simplest 'multi-panel' model, to R980 for the large 'Angolan free-tailed' model.

Important contacts

SOUTH AFRICAN BAT INTEREST GROUPS

1. Durban Bat Interest Group (DBIG)
   Chairman: Dr Peter Taylor
   Durban Natural Science Museum
   P.O. Box 4085
   Durban 4000
   Tel: 031-311 2244
   Fax: 031-311 2242
   Internet: **http://contra.biology.und.ac.za/dbig/index.htm**
   e-mail: **petert@crsu.durban.gov.za**

   DBIG: Pietermaritzburg
   Contacts: John Craigie and Michelle Hamilton
   KwaZulu-Natal Nature Conservation Service
   P.O. Box 13053
   Cascades 3202
   Tel: 033-845 1999
   Fax: 033-847 1037
   e-mail: **jcraigie@kznncs.org.za**, **mhamilton@kznncs.org.za**

2. Gauteng Bat Interest Group (GBIG)
   Secretary: Wendy Berry
   P.O. Box 710
   Irene 0062
   Tel: 012-667 2239
   e-mail: **cupar@mweb.co.za**

3. Cape BAT (Bat Action Team)
   Contact: John Taylor
   Director: Linton Projects (Pty) Ltd
   P.O. Box 603
   Rondebosch 7701
   Tel: 021-689 4406 (w)
   021-975 4621 (h)
   Fax: 021-689 4411 (w)
   021-975 1248 (h)
   Cell: 082 564 6001
   Internet: **http://www.cwr.co.za**
   e-mail: **taylorj@linton.co.za**

   or:

   Dr David Jacobs, Dr Gary Bronner
   Department of Zoology
   University of Cape Town
   Rondebosh 7700
   Tel: 021-650 4011
   Fax: 021-650 3301
   e-mail: **djacobs@botzoo.uct.ac.za**

OTHER SOUTHERN AFRICAN CENTRES OF BAT RESEARCH/ CONSERVATION

Eastern Cape

Lloyd Wingate/Fred Kigozi
Amathole Museum
P.O. Box 1434
King William's Town 5600
Tel: 0433-24506
Fax: 0433-21569
e-mail: **fred@amathole.org.za**

Prof. Ric Bernard
Department of Zoology
Rhodes University
Grahamstown 6139

Tel: 0461-318 530
Fax: 0461-24377
e-mail: **zorb@warthog.ru.ac.za**

Free State

Nico Avenant (Mammalogist)
Mammalogy Department
National Museum
P.O. Box 266
Bloemfontein 9300
Tel: 051-447 9609
Fax: 051-447 6273
e-mail: **navenant@nasmus.co.za**

Johan Watson
Department of Environment Affairs and Tourism
P.O. Box 517
Bloemfontein 9300
Tel: 051-403 3004
Fax: 051-403 3024

KwaZulu-Natal

Leon Vivier
Department of Zoology
University of Zululand
KwaDlangezwa 3886
Tel: 035-793 3911
Fax 035-793 3162
e-mail: **lvivier@pan.uzulu.ac.za**

Gauteng

Endangered Wildlife Trust (EWT)
Deputy Director: Dr Gerard Verdoorn
The Goldfields Environmental Centre
Johannesburg Zoological Gardens
Private Bag X11
Parkview 2122
Tel: 011-486 1102
Fax: 011-486 1506
Internet: **http://www.ewt.org.za/**
e-mail: **ewt@ewt.org.za**

Dr I.L. (Naas) Rautenbach
Transvaal Museum
P.O. Box 413
Pretoria 0001
Tel: 012-322 7632
Fax: 012-322 7939
e-mail: **rauten@nfi.co.za**

Dr Mac van der Merwe
Zoology Department
University of Pretoria
Pretoria 0002
Tel: 012-420 2058
Fax: 012-362 5242
e-mail: **mac@scientia.up.ac.za**

Swaziland

Dr Ara Monadjem
University of Swaziland
Private Bag 4
Kwaluseni
Swaziland
Tel: 09268-84747
Fax: 09268-85276
e-mail: **ara@science.uniswa.sz**

Namibia

Mike Griffin
Specialist Support Services
Ministry of Environment and Tourism
Private Bag 13306
Windhoek
Namibia
e-mail: **ssaurus@iafrica.co.na**

Zimbabwe

F.P.D. (Woody) Cotterill (Curator of Mammals)
Zimbabwe Natural History Museum
P.O. Box 240
Bulawayo, Zimbabwe
Tel: 09-263 9 64424/7
Fax: 09-263 9 504709
e-mail: **woodyc@gatorzw.com**

BAT HOUSES

Mr Nigel Fernsby
Birds and Bats – Environmental Products
P.O. Box 34540
Erasmia 0023

Tel/Fax: 012-659 0087
e-mail: **fernsby@netactive.co.za**

Michelle Hamilton
KwaZulu-Natal Nature Conservation Service
P.O. Box 13053
Cascades 3202
Tel: 033-845 1999
Fax: 033-847 1037
e-mail: **mhamilton@kznncs.org.za**

SOUTHERN AFRICAN SPELEOLOGICAL SOCIETIES

1. The Secretary
   Cape Peninsula Speleological Group
   P.O. Box 4812
   Cape Town 8000

2. The Secretary
   Gauteng Exploration Society
   P.O. Box 157
   Modderfontein 1645

3. Namibia
   Contact: Eugene Marais
   Entomology Centre
   National Museum of Namibia
   P.O. Box 1203
   Windhoek
   Namibia
   Tel: 09264-61-29-34437
   Internet: **http://www.natmus.cul.na/ento/index.html**
   e-mail: **insects@natmus.cul.na**

INTERNATIONAL BAT CONSERVATION SOCIETIES

1. Bat Conservation International
   Director: Dr Merlin D. Tuttle
   P.O. Box 162603
   Austin
   Texas 78716
   USA
   Internet: **http://www.batcon.org**
   e-mail: **french@batcon.org**

2. The Bat Conservation Trust
   15 Cloisters House
   8 Battersea Park Road
   London SW8 4BG
   England, UK
   Internet: **http://www.bats.org.uk**
   e-mail: **enquiries@bats.org.uk**

3. Eurobats
   Contact: Eric Blencowe
   Martin-Luther-King Str. 8
   D-53175
   Bonn
   Germany
   e-mail: **eurobats@uno.de**

4. IUCN/SSC (Species Survival Commision) Chiropera Specialist Group
   Internet: **http://www.iucn.org.uk/**

For *1996 IUCN Red List of Threatened Animals*, see the following web site:
Internet: **http://www.wcmc.org/**

Glossary

Aerial feeding — Refers to the catching of insect prey on the wing.

Ambient temperature — Outside, environmental temperature.

Amplitude — Refers to the intensity of a sound, as measured in decibels from a fixed distance (10 cm) from the bat's mouth.

Angle of attack (of wing) — Angle of rotation of bat wings around transverse axis, i.e. 'pitch'. Fast-flying bats, such as free-tailed bats, have a small angle of attack while slow-flying bats, such as horseshoe bats, have a steeper angle of attack.

Anterior — Forward end; nearest the nose of the bat.

Antitragus — A lobe of skin arising from the margin of the outer ear.

Arm wing — Part of the wing membrane between the forearm and the body of the bat. Also called the plagiopatagium. Primarily load-bearing.

Aspect ratio — The square of the wingspan, divided by the wing area. Bats with wide, short, and rounded wings have a low aspect ratio while bats with narrow, pointed wings have a high aspect ratio.

Baculum — A bone found in the penis of some mammals, including many species of bats.

Bandwidth — Difference in frequency between the minimum and maximum frequencies contained in the echolocation calls of bats. Narrow-band calls span a smaller range of frequencies than broad-band calls.

Bat detector — Electronic device which converts the ultrasonic echolocation calls of bats into an audible signal.

Bat house — Wooden, box-like structure designed for accommodating colonies of bats.

Biodiversity	The variety and variability among living things, including the number of species, number of distinct clades, within species genetic variation, and 'functional' diversity (the countless ways in which organisms function and interact with each other and the environment).
Biome	A major vegetation type.
Calcar	A bony spur projecting from the ankle of certain bats, e.g. the slit-faced bats, and supporting the edge of the tail membrane (plural calcanea).
Camber	Front to rear curvature of wings.
Canines	The longest most distinctly pointed teeth.
Carnivorous	Meat-eating; carnivorous species of bats feed on fish, frogs, rodents, and other bats.
Chiroptera	Literally meaning 'hand-wing', the name giving to the mammalian order of bats.
Classification (biological)	System of arranging organisms into groups in a hierarchical arrangement (e.g. species, genera, families, orders, classes, and kingdoms), based on similarity and evolutionary relationship.
Clutter	Refers to more closed habitats within the vegetation canopy, where free movement of bats is hindered.
Constant frequency (CF)	Refers to bat calls in which most of the sound is concentrated in a very narrow frequency band. Typical of horseshoe bats (Rhinolophidae) and leaf-nosed bats (Hipposideridae).
Decibel	Because the range of acoustic amplitudes that the ear can detect is so large, it has been convenient to define a compressed scale to describe acoustic intensities. In this decibel scale a 10-decibel (dB) difference between two sounds is perceived as a loudness difference of a factor of two.
Delayed embryonic implantation	Situation whereby fertilisation follows mating, usually just prior to winter, but the resulting blastocyst fails to implant immediately in the uterus, being delayed over the winter period until the onset of spring, whereafter implantation and normal embryonic growth continues. Known to occur, for example, in Schreibers's long-fingered bat.
Diurnal	Active by day.

Dominant frequency That frequency within the range of a bat's echolocation call where the signal is the strongest (i.e. highest amplitude or intensity). Usually corresponds to the constant frequency portion of CF calls, or the lower 'hook' of the backwards 'J' shape of some FM calls.

Doppler shift A phenomenon whereby sound produced by a moving obstacle appears to drop in frequency (pitch) as the object passes an observer (e.g. the effect you hear as a car whizzes by).

Dorsal Towards or on the back of the bat.

Draculin A pharmaceutical produced commercially from the blood of the common vampire bat (*Desmodus rotundus*) to alleviate blood clots in heart attack patients.

Duty cycle The echolocation calls of bats can be classified as high duty cycle or low duty cycle. In high duty cycle echolocation, individual calls are of longer duration than the inter-call intervals (the signal is switched 'on' most of the time), while the reverse is true for low duty cycle calls.

Duvenhage virus Name given to a lyssavirus ('rabies-like virus') described after its discovery following the death of Mr Duvenhage resulting from a bite from an apparently infected bat.

Echolocation The ability of bats and other animals to orient themselves and locate obstacles and their prey, using echoes from sounds emitted typically from the mouth or nostrils.

Endemic A species that is unique to a habitat or geographic area.

Epauletted fruit bats Bats of the fruit bat genus, *Epomophorus*, whose members possess long white shoulder hairs in adult males.

Forearm That part of the arm between the wrist and elbow. Length of the forearm is a commonly used measurement of size in bats.

Free-tailed bat Bat belonging to the Family Molossidae, in which the terminal portion of the tail projects beyond the hind margin of the tail membrane.

Frequency The 'pitch' of a sound (high or low), determined by the number of vibrations per second, measured in Hertz (1 Hertz = 1 vibration per second).

Frequency division Refers to bat detectors which transform the entire spectrum of sound by producing one pulse per predetermined number of incoming pulses, say 8 or 10, thus lowering the frequency into the audible range.

Frequency-modulated (FM) Refers to bat calls which have a non-zero bandwidth, usually sweeping from high to lower frequencies.

Frugivorous Fruit eating.

Gestation The time from fertilisation to birth of young.

Gleaning The ability of bats to snatch stationary prey from surfaces such as branches or the ground.

Hand wing The part of the wing membrane between the digits of the hand. Also called the dactylopatagium. Produces the thrust for forward movement.

Harem Group of females associating and mating with a single dominant male.

Harmonics The different frequencies of complex sounds are not randomly ordered but follow a precise sequence of related frequencies. The lowest, or first harmonic, is usually termed the fundamental; further harmonics are exact multiples of the fundamental, i.e. x2, x3, x4, etc.

Harp trap Harp-shaped device for trapping bats, originally designed by D.G. Constantine and modified by M.D. Tuttle. Two metal rectangular frames strung with fine fishing line are suspended close together with the lines offset, above a catch bag. Bats negotiating the first bank of strands collide with the second bank and slide downwards into the catch bag, which has smooth sides to prevent the bats from escaping.

Heterodyne With a heterodyne detector, the bat's ultrasound is converted into an electronic signal and compared with a variable internal signal, so that we listen to the (audible) difference between these two signals.

Hibernaculum Winter hibernating site.

Hibernation Process whereby the body metabolism is shut down to a basic level in response to winter cooling, thus allowing body temperature to fall to that of the environment, or to just above freezing in sub-zero temperatures.

Histoplasmosis (acute benign) Mild lung disease which can be caused by inhaling spores of the fungus *Histoplasma capsulatum*, living in moist bird or bat guano.

Horseshoe bats Bats of the family Rhinolophidae, whose members all possess complex structured noseleafs retaining a basic horseshoe shape of the central fold surrounding the nostrils.

Incisors The small, often tiny teeth at the front of the mouth.

Insectivorous Feeding on insects and other small invertebrates.

IUCN Acronym for International Union for the Conservation of Nature.

Kilohertz (kHz) Unit for measuring frequency.

Lactating Producing milk from mammary glands (mammae).

Lagos virus A rabies-like virus first isolated from fruit bats in the vicinity of Lagos, Nigeria.

Leading edge (of wing) Narrow part of the wing membrane in front of the arm bones. Also called the propatagium. Determines the angle of attack of the wing.

Leaf-nosed bats Bats of the family Hipposideridae, whose members all possess simple structured noseleafs having an elliptical transverse fold above the nostrils.

Lindane An organochloride, otherwise known as gamma HCH, one of the most common ingredients of pesticide fluids applied to timber, which has been shown to be highly toxic to bats and people.

Linnaean hierarchy Hierarchical system of names used to classify all living organisms, from kingdom down to species and subspecies.

Lyssavirus A group of viruses related to rabies.

Manoeuvrability Ability of bats to fly in confined or cluttered spaces without stalling or colliding with obstacles.

Maternity colony Colony of bats in which females give birth to and rear their young until they are capable of flying and feeding themselves.

Mating system System describing mating relationships between males and females, e.g. Polygyny, monogamy, or promiscuity.

Megachiroptera (megabats) The suborder of bats comprising Old World fruit-eating bats, possessing large eyes, and two claws on the wing.

Metacarpal	The long bone, closest to the wrist, in each finger.
Methyl bromide	A highly toxic gaseous fumigant, widely used in conjunction with a tent to eradicate insect pests, in particular wood borer.
Microchiroptera (microbats)	The suborder of bats comprising all the insect-eating bats, blood-feeders and animal-feeders, and New World fruit-eaters, possessing small eyes and a single claw on the wing.
Migration	Regular movement from one locality to another, usually prompted by seasonal climate changes. In bats, hibernators migrate from warm to cooler caves in winter, while non-hibernating migrators move to warmer climates in winter.
Mistnet	Fine net suspended between two poles and used to trap birds and bats.
Molars	The back teeth, larger and complex in shape with a pattern of enamel ridges and grooves for grinding food.
Monoestrus	Having a single oestrus period per year.
Monogamy	Where male and female partners form a life-long mating bond; rare in bats and has not been shown to occur in any southern African species.
Nectarivorous	Nectar eating.
Nocturnal	Active by night.
Noseleaf	System of skin flaps arranged around the nostrils of some bats, to assist with concentrating outgoing echolocation signals.
'Perch-hunt'	Method of hunting by hanging on a perch and making lightning-quick attacks when prey is detected moving past.
Phalanx	The shorter bones in the wings and toes (plural phalanges)
Pinnae	Another name for the external, cartilaginous part of the ear.
Plain-faced bats	Bats of the family Vespertilionidae, whose members lack facial ornamentation, and also have a tail completely enclosed in the tail membrane. Sometimes called evening bats or vesper bats.

Polygyny Where one dominant male associates and mates with a harem of females.

Polyoestrus Having more than one oestrus period per year.

Promiscuous Where matings occur more or less randomly in the population.

Pulse repetition rate Number of echolocation calls (pulses) per second.

Rabies Fatal disease in humans and other mammals which is spread by transmission of the virus through the bite of an infected mammal.

Retarded embryonic development Growth of the embryo is retarded during winter hibernation and only resumes after hibernation is over (e.g. Sundevall's leaf-nosed bat).

Seasonal monoestry Only a single pregnancy and birth season occurs during the restricted breeding season (e.g. the majority of bats, including most plain-faced bats).

Seasonal polyoestry More than one pregnancy and birth season occurs during the restricted breeding season (e.g. Wahlberg's epauletted fruit bat, Angolan free-tailed bat and little free-tailed bat).

Sheath-tailed bats Bats of the family Emballonuridae, whose members possess a tail which protrudes from the dorsal surface of the tail membrane.

Slit-faced bats Bats of the family Nycteridae, whose members possess a noseleaf concealed within a deep vertical opening on the face. Ear length is generally relatively longer than in any other southern African bat.

Species richness The number of species listed for unit area.

Speleology The science of cave exploration.

Sperm storage Process whereby, after mating occurs in autumn, viable sperms are stored and nourished in the female's uterus or oviducts throughout the hibernation period until spring.

Tail membrane The part of the bat's membrane enclosed by the tail and two hind limbs. Also called interfemoral membrane or uropatagium.

Taxonomy The scientific classification of species into a hierarchical system that reflects what is known about their evolutionary relationships.

Thermolability — Tendency to show tolerance for fluctuation in body temperature.

Throat patch — A circular pouch of skin on the throat. More developed in males. Found in free-tailed and sheath-tailed bats. Also called gular pouch.

Time expansion bat detector — A bat detector which digitally provides the equivalent to recording sound on a high speed tape recorder, and then playing it back at a lower speed.

Torpor — The ability of bats and other mammals to drop their body temperature in response to cooling or reduced food availability, or both. Often for short periods only.

Tragus — Specialised flap in the ear of most microbats, not found in other mammals.

Transponder — Minute transmitters which can be implanted under an animal's skin, and which must subsequently be detected at close range using a specialised scanner.

Tympanate — Refers to the presence of hearing organs termed tympani in moths.

Ultrasound — Above the frequency threshold of human hearing, usually taken as approximately 20 kHz.

Ventral — Towards or on the underside of the bat.

Wavelength — The distance between the troughs (or crests) of two successive sound waves. The shorter the wavelength, the higher the frequency.

Wing loading — Defined as the bat's body weight carried in flight, divided by its wing area. Bats with low wing loading can fly at slower speeds without stalling.

Wing pouch — Flap of skin extending from the forearm to the fifth finger forming a pocket near the wrist. Also called radiocarpal pouch. Found in sheath-tailed bats and some *Myotis* bats.

Bibliography

Acharya, L. 1992. *Epomophorus wahlbergi*. *Mammalian Species* 394: 1–4.

Aldridge, H.D. Jr. and Rautenbach, I.L. 1987. Morphology, echolocation and resource partitioning in insectivorous bats. *Journal of Animal Ecology* 56: 763–778.

Anon., 1985. *Bats in roofs: A guide for surveyors*. Nature Conservancy Council of Britain (now called English Nature), Peterborough. Available from: English Nature, Interpretative Branch, Northminster House, Peterborough PE1 1UA, UK.

Baillie, J. and Groombridge, B. 1996. *1996 IUCN Red List of Threatened Animals*. IUCN, Gland, Switzerland and Cambridge, UK.

Barnard, S.M. 1995. *Bats in captivity*. Wild Ones Animal Books, Springville, California.

Bergmans, W. 1988. Taxonomy and biogeography of African fruit bats (Mammalia, Megachiroptera). 1. General introduction; material and methods; results; the genus *Epomophorus* Bennett, 1836. *Beaufortia* 38: 75–146.

Bergmans, W. 1989. Taxonomy and biogeography of African fruit bats (Mammalia, Megachiroptera). 2. The genera *Micropterus* Matschie, 1899, *Epomops* Gray, 1870, *Hypsignathus* H. Allen, 1861, *Nanonycteris* Matschie, 1899, and *Plerotes* Andersen, 1910. *Beaufortia* 39: 89–153.

Bergmans, W. 1990. Taxonomy and biogeography of African fruit bats (Mammalia, Megachiroptera). 3. The genera *Scotonycteris* Matschie, 1894, *Casinycteris* Thomas, 1910, *Pteropus* Brisson, 1762, and *Eidolon* Rafinesque, 1815. *Beaufortia* 40: 111–177.

Bergmans, W. 1994. Taxonomy and biogeography of African fruit bats (Mammalia, Megachiroptera). 4. The genus *Rousettus* Gray, 1821. *Beaufortia* 44: 79–126.

Bergmans, W. 1997. Taxonomy and biogeography of African fruit bats (Mammalia, Megachiroptera). 5. The genera *Lissonycteris* Andersen, 1912, *Myonycteris* Matschie, 1899 and *Megaloglossus* Pagenstechner, 1885; general remarks and conclusions; annex; key to all species. *Beaufortia* 47: 11–90.

Bernard, R.T.F. 1980. Reproductive cycles of *Miniopterus schreibersii natalensis* (Kuhl, 1819) and *M. fraterculus* (Thomas & Schwann, 1906). *Annals of the Transvaal Museum* 32: 55–64.

Bernard, R.T.F. 1982a. Female reproductive cycle of *Nycteris thebaica* (Microchiroptera) from Natal, South Africa. *Zeitschrift für Säugetierkunde* 47: 12–18.

Bernard, R.T.F. 1982b. Monthly changes in the female reproductive organs and the reproductive cycle of *Myotis tricolor* (Vespertilionidae: Chiroptera). *South African Journal of Zoology* 17: 79–84.

Bernard, R.T.F. 1983. Reproduction of *Rhinolophus clivosus* (Microchiroptera) in Natal, South Africa. *Zeitschrift für Säugetierkunde* 48: 321–329.

Bernard, R.T.F. 1985. Reproduction in the Cape horseshoe bat (*Rhinolophus capensis*) from South Africa. *South African Journal of Zoology* 20: 129–135.

Bernard, R.T.F. 1986. The natural history of some eastern Cape bats. III. Reproduction and development. *The Naturalist* 30(3): 55–56.

Bernard, R.T.F. 1989. The adaptive significance of reproductive delay phenomena in some South African Microchiroptera. *Mammal Review* 19(1): 27–34.

Bernard, R.T.F. 1994. Reproductive synchrony and annual variation in foetal growth rate in the long-fingered bat (*Miniopterus schreibersii*). *Journal of Zoology, London* 232: 485–490.

Bernard, R.T.F. and Bester, G.J. 1988. Roost selection by the long-fingered bat, *Miniopterus schreibersii*, is influenced by reproductive condition, and roost microclimate and structure. *South African Journal of Science* 84: 921–922.

Bernard, R.T.F. and Cumming, G.S. 1997. African bats: Evolution of reproductive patterns and delays. *Quarterly Review of Biology* 72(3): 253–274.

Bernard, R.T.F. and Meester, J.A.J. 1982. Female reproduction and the female reproductive cycle of *Hipposideros caffer caffer* (Sundevall, 1846) in Natal, South Africa. *Annals of the Transvaal Museum* 33: 131–144.

Bernard, R.T.F., Cotterill, F.P.D. and Fergusson, R.A. 1996. On the occurrence of a short period of delayed implantation in Schreibers's long-fingered bat (*Miniopterus schreibersii*) from a tropical latitude in Zimbabwe. *Journal of Zoology, London* 238: 13–22.

Bernard, R.T.F., Happold, D.C.D. and Happold, M. 1997. Sperm storage in a seasonally reproducing African vespertilionid, the banana bat (*Pipistrellus nanus*) from Malawi. *Journal of Zoology, London* 241: 161–174.

Bouchard, S. 1998. *Chaerephon pumilus. Mammalian Species* 547: 1–6.

Boulay, M.C. and Robbins, C.B. 1989. *Epomophorus gambianus. Mammalian Species* 344: 1–5.

Bowie, R.C., Jacobs, D.S. and Taylor, P.J. 1999. Resource use by two morphologically similar insectivorous bats (*Nycteris thebaica* and *Hipposideros caffer*). *South African Journal of Zoology* 34: 27–33.

Briggs, B. and King, D. 1998. *The bat detective: A field guide for bat detection.* Stag Electronics, Shoreham-by-Sea. Note: A compact disk with recordings of the calls of British bats accompanies his book. Book and CD available from Stag Electronics, 4 Esprit Court, New Road, Shoreham-by-Sea, West Sussex, BN43 6RB, UK.

Bronner, G.N., Maloney, S.K. and Buffenstein, R. 1999. Survival tactics within thermally-challenging roosts: Heat tolerance and cold sensitivity in the Angolan free-tailed bat, *Mops condylurus. South African Journal of Zoology* 34: 1–10.

Brown, K.M. and Dunlop, J. 1997. *Rhinolophus landeri. Mammalian Species* 567: 1–4.

Campbell, C.A.R. 1925. *Bats, mosquitoes and dollars.* The Stratford company, Boston, USA.

Catto, C. 1994. *Bat detector manual*. The Bat Conservation Trust, London, UK.

Churchill, S. Personal communication. 26 Riverview Drive, Paradise, South Australia, Australia.

Churchill, S. 1998. *Australian bats*. New Holland, Sydney, Australia.

Churchill, S., Draper, R. and Marais, E. 1997. Cave utilisation by Namibia bats. *South African Journal of Wildlife Research* 27(2): 44–50.

Cotterill, F.P.D. Personal communication. Curator of Mammals, Zimbabwe Natural History Museum, P.O. Box 240, Bulawayo, Zimbabwe.

Cotterill, F.P.D. 1995. New fruit bat species for southern Africa. *Zimbabwe Wildlife* January–March: 11.

Cotterill, F.P.D. 1996a. New distribution records of insectivorous bats of the families Nycteridae, Rhinolophidae and Vespertilionidae (Microchiroptera: Mammalia) in Zimbabwe. *Arnoldia Zimbabwe* 10(8): 71–89.

Cotterill, F.P.D. 1996b. New distribution records of free-tailed bats (Microchiroptera: Molossidae) in Zimbabwe. *Arnoldia Zimbabwe* 10(9): 91–102.

Cotterill, F.P.D. 1998. Female reproduction in two species of tropical horseshoe bats (Rhinolophidae) in Zimbabwe. *Journal of Mammalogy* 79(4): 1306–1316.

Cotterill, F.P.D. In press (a). The first known occurrence of Thomas's flat-headed bat, *Mimitellus moloneyi thomasi* (Microchiroptera: Mammalia) in southern Africa from Mozambique. *Arnoldia Zimbabwe*.

Cotterill, F.P.D. In press (b). Further notes on the natural history of some large Afrotropical free-tailed bats (Molossidae: Tadarida). *Arnoldia Zimbabwe*.

Cotterill, F.P.D. In press (c). New distribution records of leaf-nosed bats (Microchiroptera: Hipposideridae) in Zimbabwe. *Arnoldia Zimbabwe*.

Cotterill, F.P.D. In press (d). Natural history notes on some African bats, with a critique of contemporary approaches to chiropteran taxonomy. *Arnoldia Zimbabwe*.

Cotterill, F.P.D. In press (e). Distributional and taxonomic notes on some Afrotropical horseshoe bats (Microchiroptera: Rhinolophidae) with a diagnostic key. *Arnoldia Zimbabwe*.

Cotterill, F.P.D. In press (f). New specimens of lesser house bats (Vespertilionidae: Scotoecus) from Mozambique and Zambia. *Arnoldia Zimbabwe*.

Cotterill, F.P.D. and Fergusson, R.A. 1993a. Capturing free-tailed bats (Chiroptera: Molossidae): The description of a new trapping device. *Journal of Zoology, London* 231: 645–651.

Cotterill, F.P.D. and Fergusson, R.A. 1993b. Seasonally polyestrous reproduction in a free-tailed bat *Tadarida fulminans* (Microchrioptera: Molossidae) in Zimbabwe. *Biotropica* 25(4): 487–492.

Cotterill, F.P.D. and Fergusson, R.A. 1999. Reproductive ecology of Commerson's leaf-nosed bats *Hipposideros commersoni* (Chiroptera: Hipposideridae) in south-central Africa: Interactions between seasonality and large body size; and implications for conservation. *South African Journal of Zoology* 34(1): 53–63.

Cumming, G.S. and Bernard, R.T.F. 1997. Rainfall, food abundance and timing of parturition in African bats. *Oecologia* 111: 309–317.

Dengis, C.A. 1996. *Taphozous mauritianus*. *Mammalian Species* 522: 1–5.

Dunlop, J. 1997. *Coleura afra. Mammalian Species* 566: 1–4.

Fenton, M.B. Personal communication. Department of Biology, York University, North York, Ontario, M3J 1P3, Canada.

Fenton, M.B. 1985. The feeding behaviour of insectivorous bats: Echolocation, foraging strategies, and resource partitioning. *Transvaal Museum Bulletin* 21: 5–19.

Fenton, M.B. 1986. *Hipposideros caffer* (Chiroptera: Hipposideridae) in Zimbabwe: Morphology and echolocation calls. *Journal of Zoology, London* 210: 347–353.

Fenton, M.B. 1992a. *Bats*. Facts on File, New York, USA.

Fenton, M.B. 1992b. Pelage patterns and crypsis in roosting bats: *Taphozous mauritianus* and *Epomophorus* species. *Koedoe* 35(2): 49–55.

Fenton, M.B. and Bell, G.P. 1981. Recognition of species of insectivorous bats by their echolocation calls. *Journal of Mammalogy* 62(2): 233–243.

Fenton, M.B. and Griffin, D.R. 1997. High-altitude pursuit of insects by echolocating bats. *Journal of Mammalogy* 78(1): 247–250.

Fenton, M.B. and Rautenbach, I.L. 1986. A comparison of the roosting and foraging behaviour of three species of African insectivorous bats (Rhinolophidae, Verspertilionidae and Molossidae). *Canadian Journal of Zoology* 64: 2860–2866.

Fenton, M.B. and Thomas, D.W. 1980. Dry-season overlap inactivity patterns, habitat use, and prey selection by sympatric African insectivorous bats. *Biotropica* 12(2): 81–90.

Fenton, M.B., Bell, G.P. and Thomas, D.W. 1980. Echolocation and feeding behaviour of *Taphozous mauritianus* (Chiroptera: Emballonuridae). *Canadian Journal of Zoology* 58: 1774–1777.

Fenton, M.B., Boyle, N.G.H., Harrison, T.M. and Oxley, D.J. 1977. Activity patterns, habitat use and prey selection by some African insectivorous bats. *Biotropica* 9(2): 73–85.

Fenton, M.B., Brigham, R.M., Mills, A.M. and Rautenbach, I.L. 1985. The roosting and foraging areas of *Epomophorus wahlbergi* (Pteropodidae) and *Scotophilus viridis* (Vespertilionidae) in the Kruger National Park. *Journal of Mammalogy* 66: 461–468.

Fenton, M.B., Cumming, D.H.M. and Oxley, D.J. 1978. Prey of bat hawks and availability of bats. *Condor* 79: 495–497.

Fenton, M.B., Cumming, D.H.M., Hutton, J.M. and Swanepoel, C.M. 1987. Foraging and habitat use by *Nycteris grandis* (Chiroptera: Nycteridae) in Zimbabwe. *Journal of Zoology, London* 211: 709–716.

Fenton, M.B., Gaudet, C.L. and Leonard, M.L. 1983. Feeding behaviour of the bats *Nycteris grandis* and *Nycteris thebaica* (Nycteridae) in captivity. *Journal of Zoology, London* 200: 347–354.

Fenton, M.B., Portfors, C.V., Rautenbach, I.L. and Waterman, J.M. 1998. Compromises: Sound frequencies used in echolocation by aerial-feeding bats. *Canadian Journal of Zoology* 76: 1174–1182.

Fenton, M.B., Rautenbach, I.L., Chipese, D., Cumming, M.B., Musgrave, M.K., Taylor, J.S. and Volpers, T. 1993. Variation in foraging behaviour, habitat use, and diet of large slit-faced bats (*Nycteris grandis*). *Zeitschrift für Säugetierkunde* 58: 65–74.

Fenton, M.B., Rautenbach, I.L., Smith, S.E., Swanepoel, C.M., Grosell, J. and Van Jaarsveld, J. 1994. Raptors and bats: Threats and opportunities. *Animal Behaviour* 48: 9–18.

Fenton, M.B., Swanepoel, C.M., Brigham, R.M., Cebek, J.E. and Hickey, M.B.C. 1990. Foraging behaviour and prey selection by large slit-faced bats (*Nycteris grandis*; Chiroptera: Nycteridae). *Biotropica* 22: 2–8.

Fenton, M.B., Thomas, D.W. and Sasseen, R. 1981. *Nycteris grandis* (Nycteridae): An African carnivorous bat. *Journal of Zoology, London* 194: 461–465.

Fernsby, N. 1999. The siting, use, management and maintenance of bat houses. Unpublished report, available from the author.

Francis, C.M. 1989. A comparison of mist nets and two designs of harp traps for capturing bats. *Journal of Mammalogy* 70: 865–870.

Gelderblom, C.M., Bronner, G.N., Lombard, A.T. and Taylor, P.J. 1995. Patterns of distribution and current protection status of the Carnivora, Chiroptera and Insectivora in South Africa. *South African Journal of Zoology* 30(3): 103–114.

Griffin, D.G. 1958. *Listening in the dark*. Yale University Press, New Haven. Reprinted 1986 by Cornell University Press, Ithaca and London, USA and UK.

Griffin, D.G., Webster, R.A. and Michael, C.R. 1960. The echolocation of flying insects by bats. *Animal Behaviour* 8: 141–154.

Grzimek, B. 1990. *Multimedia Encyclopedia of Mammals*. McGraw Hill, London (available on CD-ROM).

Harrison, D.L. Personal communication. Harrison Zoological Institute, Centre for Systematics and Biodiversity Research, Bowerwood House, 15 St Botolphs Road, Sevenoaks, Kent, TN13 3AQ, UK.

Herselman, J.C. and Norton P.M. 1985. The distribution and status of bats (Mammalia: Chiroptera) in the Cape Province. *Annals of the Cape Provincial Museums (Natural History)* 16(4): 73–126.

Hill, J.E. 1974a. A review of *Laephotis* Thomas, 1901 (Chiroptera: Vespertilionidae). *Bulletin of the British Museum of Natural History (Zoology)* 27: 73–82.

Hill, J.E. 1974b. A review of *Scotoecus* Thomas, 1901 (Chiroptera: Vespertilionidae). *Bulletin of the British Museum of Natural History (Zoology)* 27: 169–188.

Hutson, A.M. 1986. *Bats in houses*. Fauna and Flora Preservation Society, London. On sale from: FFPS, c/o Zoological Society of London, Regents Park, London NW1 4RY, UK.

Hutton, J.M. 1986. The status and distribution of bats in Zimbabwe. *Cimbebasia* 8: 219–236.

Jacobs, D.S. Personal communication. Department of Botany and Zoology, University of Cape Town, Rondebosch 7700, South Africa.

Jacobs, D.S. 1999. Intraspecific variation in wingspan and echolocation call flexibility might explain the use of different habitats by the insectivorous bat,

Miniopterus schreibersii (Vespertilionidae: Miniopterinae). *Acta Chiropterologica* 1(1): 93–104.

Jacobs, D.S. and Fenton, M.B. In press. Chaerephon pumilus and Sauromys petrophilus (Chiroptera: Molossidae) in southern Africa. *African Zoology.*

Jacobsen, N.H.G. and Du Plessis, E. 1976. Observations on the ecology and biology of the Cape fruit bat *Rousettus aegyptiacus leachi* in the Eastern Transvaal. *South African Journal of Science* 72: 270–273.

Kearney, T. and Taylor, P.J. 1997. New distribution records of bats in KwaZulu-Natal. *Durban Museum Novitates* 22: 53–56.

Kunz, T.H. 1988. *Ecological and Behavioural Methods for the Study of Bats.* Smithsonian Institution Press, Washington DC, USA.

Laycock, P.A. 1976. A study of cave-dwelling Microchiroptera in the Natal midlands. MSc thesis, University of Natal, Pietermaritzburg.

Lollar, A. and Schmidt-French, B. 1998. *Captive Care and Medical Reference for the Rehabilitation of Insectivorous Bats.* Bat World Publications, Mineral Wells, Texas: 329 pp.

Long, R., Simpson, T., Ding, J., Heydon, S. and Reil, R. 1998. Bats feed on crop pests in Sacramento Valley. *California Agriculture* 52: 8–10.

Lynch, C.D. 1983. The mammals of the Orange Free State. *Memoirs van die Nasionale Museum, Bloemfontein* 18: 1–218.

Lynch, C.D. 1985. Mammalian distribution patterns in the Orange Free State. *Navorsinge van die Nasionale Museum, Bloemfontein* 4(16): 473–500.

Lynch, C.D. 1989. The mammals of the north-eastern Cape Province. *Memoirs van die Nasionale Museum, Bloemfontein* 25: 1–116.

Lynch, C.D. 1994. The mammals of Lesotho. *Navorsinge van die Nasionale Museum, Bloemfontein* 10(4): 177–241.

McCracken, G.F. 1996. Bats aloft. *Bats* 14 (3): 7–10. *Bats* magazine available from Bat Conservation International, P.O. Box 162603, Austin, Texas 78716, or can be ordered through their website: **http//www.batcon.org**

McDonald, J.T., Rautenbach, I.L. and Nel, J.A.J. 1990a. Foraging ecology of bats observed at De Hoop Provincial Nature Reserve, southern Cape Province. *South African Journal of Wildlife Research* 20(4): 133–145.

McDonald, J.T., Rautenbach, I.L. and Nel, J.A.J. 1990b. Roosting requirements and behaviour of five bat species at De Hoop Guano Cave, southern Cape Province of South Africa. *South African Journal of Wildlife Research* 20(4): 157–161.

McWilliam, A.N. 1987. The reproduction and social biology of *Coleura afra* in a seasonal environment. In *Recent advances in the study of bats,* eds. M.B. Fenton, P. Racey and J.M.V. Rayner, Cambridge University Press, Cambridge, UK: 324–350.

Mitchell-Jones, A.J. and McLeish, A.P. 1999. *Bat Worker's Manual,* 2nd edition. English Nature, Peterborough. Available from: English Nature, Interpretative Branch, Northminster House, Peterborough PE1 1UA, UK.

Mitchell-Jones, A.J., Hutson, A.M. and Racey, P.A. 1993. The growth and development of bat conservation in Britain. *Mammal Review* 23(3/4): 139–148.

Monadjem, A. 1998a. *Mammals of Swaziland*. The Conservation Trust of Swaziland and Big Game Parks.

Monadjem, A. 1998b. Reproduction in the little free-tailed bat *Chaerephon pumila* (Microchiroptera: Molossidae) in Swaziland. *Durban Museum Novitates* 23: 39–41.

Nowak, R.M. and Paradiso, J.L. 1983. *Walker's mammals of the world*, 4th edition. The John Hopkins University Press, Baltimore, Maryland, USA.

Obrist, M., Aldridge, H.D.J.N. and Fenton, M.B. 1989. Roosting and echolocation behaviour of the African bat, *Chalinolobus variegatus*. *Journal of Mammalogy* 70(4): 828–833.

Pearl, D.L. 1994. *Rhinolophus hildebrandti*. *Mammalian Species* 486: 1–3.

Rautenbach, I.L. 1982. Mammals of the Transvaal. *Ecoplan Monograph* 1: 1–211.

Rautenbach, I.L. and Fenton, M.B. 1992. Bats from Mana Pools National Park in Zimbabwe and the first record of *Eptesicus rendallii* from the country. *Zeitschrift für Säugetierkunde* 57: 112–115.

Rautenbach, I.L. and Nel, J.A.J. 1975. Further records of smaller mammals from the Kalahari Gemsbok National Park. *Koedoe* 18: 195–198.

Rautenbach, I.L. and Nel, J.A.J. 1978. Three species of microchiropteran bats recorded for the first time from the southwest Cape Biotic Zone. *Annals of the Transvaal Museum* 31: 157–161.

Rautenbach, I.L., Bronner, G.N. and Schlitter, D.A. 1993. Karyotypic data and attendant systematic implications for the bats of southern Africa. *Koedoe* 36(2): 87–104.

Rautenbach, I.L., Fenton, M.B. and Braack, L.E.O. 1985. First records of five species of insectivorous bats from the Kruger National Park. *Koedoe* 28: 73–80.

Rautenbach, I.L., Schlitter, D.A. and Braack, L.E.O. 1984. New distributional records of bats for the Republic of South Africa, with special reference to the Kruger National Park. *Koedoe* 27: 131–135.

Richardson, E.J. 1997. *Manual on bat care and rehabilitation*. Available from Durban Bat Interest Group, P.O. Box 4085, Durban 4000.

Richardson, E.J. and Taylor, P.J. 1995. New observations on the large-eared free-tailed bat *Otomops martiensseni* in Durban, South Africa. *Durban Museum Novitates* 20: 72–74.

Roberts, A. 1951. *Mammals of South Africa*. Central News Agency, Cape Town.

Robertson, J. 1990. *The Complete Bat*. Chatto & Windus, London, UK.

Rosevear, D.R. 1965. *The bats of West Africa*. British Museum (Natural History), London, UK.

Sapsford, C. Personal communication. 29 Tuin Street, Moorreesburg, Western Cape, South Africa.

Seamark, E. Personal communication. Gauteng Nature Conservation, P.O. Box 59143, Karen Park 0118, South Africa.

Setzer, H.W. 1971. New bats of the genus *Laephotis* from Africa (Mammalia: Chiroptera). *Proceedings of the Biological Society of Washington* 84(32): 259–264.

Shortridge, G.C. 1934. *The mammals of South West Africa*, Volumes I and II. Heinemann, London, UK.

Skinner, J.D. and Smithers, R.H.N. 1990. *The mammals of the Southern African subregion*. University of Pretoria, Pretoria.

Smithers, R.H.N. 1971. The mammals of Botswana. *Museum Memoirs of the National Museums and Monuments, Rhodesia* 4: 1–340.

Smithers, R.H.N. 1983. *The mammals of the southern African subregion*. University of Pretoria, Pretoria.

Smithers, R.H.N. 1986. South African Red Data Book: Terrestrial mammals. *South African National Scientific Programmes Report* No. 125: 1–216.

Smithers, R.H.N. and Tello, J.L.P.L. 1976. Checklist and atlas of the mammals of Mozambique. *Museum Memoirs of the National Museums and Monuments, Rhodesia* 8: 1–184.

Smithers, R.H.N. and Wilson, V.J. 1979. Checklist and atlas of the mammals of Zimbabwe Rhodesia. *Museum Memoirs of the National Museums and Monuments, Rhodesia* 9: 1–147.

Stebbings, R.E. 1988. *Conservation of European bats*. Christopher Helm, London, UK.

Stuart, C. and T. 1992. *Field Guide Mammals of Southern Africa*, Fourth impression. Struik, Cape Town.

Taylor, P.J. 1997. Flitting into the danger zone. *African Wildlife* 51:19–20.

Taylor, P.J. 1998a. The flying squad. *Farmers Weekly*, June 12: 22–23.

Taylor, P.J. 1998b. *The smaller mammals of KwaZulu-Natal*. University of Natal Press, Pietermaritzburg.

Taylor, P.J. 1999a. Echolocation calls of twenty southern African bat species. *South African Journal of Zoology* 34(3): 114–125.

Taylor, P.J. 1999b. Problems with the identification of southern African *Chaerephon* (Molossidae), and the possibility of a cryptic species from South Africa and Swaziland. *Acta Chiropterologica* 1(2): 191–200.

Taylor, P.J. 1999c. The role of amateurs in the growth of bat conservation and research in South Africa. *South African Journal of Zoology* 34(1): 19–26.

Taylor, P.J. and Richardson, E. 1998. *The South African Bat Workers' Manual*, Revised edition. Durban Bat Interest Group, P.O. Box 4085, Durban 4000.

Taylor, P.J. and Van der Merwe, M. 1998. Taxonomic notes on dark-winged house bats of the genus Scotoecus Thomas 1901, in Malawi. *Durban Museum Novitates* 23: 64–66.

Tidemann, C.R. and Woodside, D.P. 1978. A collapsible bat-trap and a comparison of results obtained with the trap and mist nets. *Australian Wildlife Research* 5: 355–362.

Tudge, C. 1994. Going bats over conservation. *New Scientist*, 12 March: 27–31.

Tuttle, M.D. Personal communication. Executive Director, Bat Conservation International, P.O. Box 162603, Austin, Texas 78716, USA.

Tuttle, M.D. 1988. *America's neighbourhood bats: Understanding and learning to live in harmony with them*. University of Texas, Austin. Revised 1997. Note: this publication has an excellent section under the heading, Suggested reading, which summarises 14 studies worldwide documenting the impact of bats on pest insects.

Tuttle, M.D. and Hensley, D.L. 1993. *The bat house builder's handbook*. Bat Conservation International, Austin, Texas. Revised 1997. On sale from BCI, P.O. Box 162603, Austin, Texas 78716, or can be ordered through their website: **http//www.batcon.org**

Tuttle, M.D. and Taylor, D.A.R. 1994. Bats and mines. *Bat Conservation International Resource Publication* 3: 1–41.

Van der Merwe, M. 1973a. Aspects of hibernation and winter activity of the Natal clinging bat, *Miniopterus schreibersii natalensis* (A. Smith, 1834) on the Transvaal Highveld. *South African Journal of Science* 69: 116–118.

Van der Merwe, M. 1973b. Aspects of social behaviour of the Natal clinging bat, *Miniopterus schreibersii natalensis* (A. Smith, 1834). *Mammalia* 37: 380–389.

Van der Merwe, M. 1973c. Aspects of temperature and humidity in preferred hibernation sites of the Natal clinging bat, *Miniopterus schreibersii natalensis* (A. Smith, 1834). *Zoologica Africana* 8: 121–134.

Van der Merwe, M. 1975. Preliminary study on the annual movements of the Natal clinging bat. *South African Journal of Science* 71: 237–241.

Van der Merwe, M. 1978. Postnatal development and mother-infant relationships in the Natal clinging bat, *Miniopterus schreibersii natalensis* (A. Smith, 1834). In *Proceedings of the 14th International Bat Research Conference*, Kenya, National Academy of Arts and Sciences: 309–322.

Van der Merwe, M. 1979. Foetal growth curves and seasonal breeding in the Natal clinging bat, *Miniopterus schreibersii natalensis*. *South African Journal of Zoology* 14: 17–21.

Van der Merwe, M. 1980. Importance of *Miniopterus schreibersii natalensis* in the diet of barn owls. *South African Journal of Wildlife Research* 10: 15–17.

Van der Merwe, M. 1982. Bat vectors of rabies. *South African Journal of Science* 78: 421–422.

Van der Merwe, M. 1989. Longevity in Schreibers's long-fingered bat. *South African Journal of Wildlife Research* 19(2): 87–89.

Van der Merwe, M. 1990. Polytocy in the Cape serotine bat *Eptesicus capensis* (A. Smith 1829) from the southern African subregion. *South African Journal of Zoology* 25: 199–203.

Van der Merwe, M. 1994. Reproductive biology of the Cape serotine bat, *Eptesicus capensis* in the Transvaal, South Africa. *South African Journal of Zoology* 29(1): 36–39.

Van der Merwe, M. and Rautenbach, I.L. 1986. Multiple births in Schlieffen's bat, *Nycticeius schlieffenii* (Peters, 1859) (Chiroptera: vespertilionidae) from the southern African subregion. *South African Journal of Zoology* 21: 48–50.

Van der Merwe, M. and Rautenbach, I.L. 1987. Reproduction in Schlieffen's bat, *Nycticeius schlieffenii*, in the eastern Transvaal lowveld, South Africa. *Journal of Reproductive Fertility* 81: 41–50.

Van der Merwe, M. and Rautenbach, I.L. 1990. Reproduction in the rusty bat, *Pipistrellus rusticus*, in the northern Transvaal bushveld, South Africa. *Journal of Reproductive Fertility* 89: 537–542.

Van der Merwe, M., Giddings, S.R. and Rautenbach, I.L. 1987. Post-partum oestrus in the little free-tailed bat, *Tadarida (Chaerephon) pumila* (Microchiroptera: Molossidae) at 24° S. *Journal of Zoology, London* 213: 317–326.

Van der Merwe, N.J., Rautenbach, I.L. and Penzhorn, B.L. 1988. A new pattern of early embryonic development in the seasonally breeding non-hibernating lesser yellow house bat, *Scotophilus borbonicus* (E. Geoffroy, 1803) (Chiroptera: Vespertilionidae). *Annals of the Transvaal Museum* 34: 152–156.

Van der Merwe, M., Rautenbach, I.L. and Van der Colf, W.J. 1986. Reproduction in females of the little free-tailed bat, *Tadarida (Chaerephon) pumila*, in the eastern Transvaal, South Africa. *Journal of Reproductive Fertility* 77: 355–364.

Volpers, T. and Kumirai, A. 1995. Ecological aspects of the distribution of *Epomophorus*-fruit bats (Mammal., Chiropt.) in Zimbabwe. *Verhandl. Ges. f. Ökol.* 24: 89–92.

Volpers, T. and Kumirai, A. 1996. A review of the genus *Epomophorus* Bennett, 1836 (Mammalia, Megachiroptera) in Zimbabwe. *Arnoldia Zimbabwe* 10(10): 103–109.

Watson, J.P. 1990a. New distribution records of *Laephotis* in South Africa and Lesotho. *Navorsinge van die Nasionale Museum, Bloemfontein* 7(4): 61–70.

Watson, J.P. 1990b. Westward range extension of Temminck's hairy bat in South Africa and Lesotho. *South African Journal of Wildlife Research* 20(3): 119–120.

Watson, J.P. 1998. New distributional records for three microchiropteran bats (Vespertilionidae, Rhinolophidae) from the Free State Province, South Africa. *South African Journal of Wildlife Research* 28(4): 127–131.

Whitaker, J.O. Jr. 1993. Bats, beetles and bugs. *Bats* 11: 23.

Whitaker, J.O. Jr., Neefus, C., and Kunz, T.H. 1996. Dietary variation in the Mexican free-tailed bat (*Tadarida brasiliensis mexicana*). *Journal of Mammalogy* 77(3): 716–724.

Williams-Whitmer, L.M. and Brittingham, M.C. 1995. *A homeowner's guide to Northeastern bats and bat problems*. The Pennsylvania State University, University Park. Available from: Publication Distribution Center, The Pennsylvania State University, 119 Agricultural Administration Building, University Park, PA 16802, USA. Note: A video, 'Bat-Free Belfries: A Guide to Bat-proofing' (23 minutes), is also available at a cost of $US35 from the same publisher.

Index